MY PLASTIC -FREE HOME

Kate Jones

MY PLASTIC -FREE HOME

Simple Steps to Live More Sustainably

HarperCollins*Publishers*

CONTENTS

INTRODUCTION	7
MY PLASTIC-FREE...	
... KITCHEN	12
... BATHROOM	74
... CLEANING	113
... LITTLE LUXURIES	140
... DECOR	186
... FAMILY FUN	210
... GARDEN	236
RESOURCES	252
PICTURE CREDITS	255

MY PLASTIC-FREE ...

INTRODUCTION

Welcome to My Plastic-Free Home, a happy little place that I have cultivated over the years using low-waste recipes, tips and tricks. My Plastic-Free Home was created from a seed that was sown when my daughter was born. We prepared for her arrival, as many new parents do, taking note of all the never-ending baby lists and everything that we 'needed' for her. By the time she entered our world I was so completely and utterly tired of all the consumerism and waste in my life, and having a beautiful new baby seemed to put everything into perspective. I realised she didn't need all of the fancy gifts and products that she was adorned with before her arrival – all she needed was love and feeding. Watching her grow confirmed this for me; nothing interested her more than our long walks in nature and the enjoyment that came from watching the seasons change, and nothing settled her more than a breath of fresh air or a change of scenery. This little person taught me how to see the world from a fresh perspective and helped me to realise how important it is to protect it.

When I was first approached to write this book, I didn't think that I had enough tips to fill the pages – but I was wrong! As soon as I put pen to paper, the ideas kept flowing and I found it such an enjoyable process. At a time where climate news can make us all feel overwhelmed and anxious, I hope that this book feels like a reassuring hug. There are so many things that we can do as individuals to make little positive changes in the world. In the time that I have been practising and sharing these steps online, I have been reassured by how people are embracing these plastic-free choices wherever they can. If we all continue to adopt a lifestyle that produces a little less waste, imagine how different our world could look.

I hope you find the suggestions in the seven sections that follow helpful and attainable. Some are quick, simple fixes; others take a little more time, but they are very much worth the effort. My hope is that you use this book in any way you can to aid your plastic-free journey, and that it also brings you a sense of comfort in the knowledge that the small things really can make a difference.

... INTRODUCTION

THE FIVE Rs OF ZERO WASTE

I couldn't write a book about being more mindful about your consumption and avoiding plastic without mentioning the five Rs of zero waste. Simply put, these are a hierarchy of principles designed to minimise waste and promote sustainability.

They are:

1 **Refuse:** avoid accepting or purchasing items that you feel are unnecessary or contribute to waste. This includes saying no to single-use plastics, as well as promotional materials or items with excessive packaging. Stop and think: do I actually need this thing I'm about to buy or is being offered to me? If you don't, just say no!

2 **Reduce**: try to cut down on the amount of waste that you generate by choosing products with minimal or no packaging. Also focus on simplifying and decluttering your life to reduce your overall consumption. In recent decades we have been sold this idea that we can afford everything that we could possibly want, but, in reality, we don't need half the things we end up buying, and our planet simply can't afford it.

3 **Reuse**: Opt for reusable items instead of disposable ones, for example, cloth bags, reusable bottles and containers. You don't have to buy all these, and you probably have lots of them in your home that you can already reuse. Reusing, repurposing or repairing items rather than discarding them is extremely helpful.

4 **Recycle**: properly sorting and recycling materials according to your local guidelines is so important to help manage waste.

5 **Rot**: I actually think this should be higher up the list because composting is game-changing for our planet. Composting waste like food scraps and garden waste turns biodegradable materials into nutrient-rich compost, which can then be used to enrich your soil while also reducing the amount of waste that ends up in landfill.

MY
PLASTIC-FREE
KITCHEN

ECO KITCHEN TIPS

When I think about trying to reduce waste around the home, the first place that comes to mind is the kitchen. I find the saying 'the kitchen is the heart of the home' to be so true. As a family we spend most of our time in this space and with two young children it keeps us very busy!

I try my best to be as eco-friendly as possible in the kitchen, and this works out most of the time. We are all aware that when life gets busy, eco-friendly habits get replaced by convenience habits, which tend to bring all of that packaging that we try to avoid – as well as ultra-processed foods, which we know aren't healthy for ourselves or our families. Over the years I have managed to create easy processes and routines that take only a few minutes, but help me to avoid those 'convenience habits' and allow me to feel prepared for the week ahead. What follows is some of those processes, as well as other simple changes you can try too, to avoid using plastics wherever possible. Both approaches combined have made my kitchen a manageable and low-waste space.

It's good to think of the following tips as a reset for you and your kitchen. I find the act of doing these tasks helps me to slow down, and making them part of my lifestyle not only makes my routine more eco-friendly, but also more mindful and enjoyable.

BUYING LOCAL, SEASONAL AND PLASTIC-FREE

I like to try and buy my food from the local grocery shops near me and focus my recipes on seasonal produce. Buying local and seasonal produce is important for so many reasons, from improving your health and supporting your local economy, to protecting the environment and promoting sustainability. It fosters a more intimate connection with your food and helps to build stronger, more resilient communities.

Fresh produce also means it typically retains more of its nutrients, as fewer preservatives and chemicals are required to extend its shelf life. Limiting transportation of food reduces greenhouse gas emissions, as well as the need for plastic storage containers.

Buying direct from farmers and local markets will create jobs and stimulate economic growth in the area where you live, and it can also help to protect nearby green spaces. If the land is being used to generate local income, it is far less likely to be sold for building purposes.

Eating seasonally means consuming produce at the time it is naturally harvested. This encourages us to have a more varied diet throughout the year, which leads to a more balanced intake of nutrients. I also find that eating seasonally means that I look forward to produce coming into season (for me, there really is nothing like the taste of a strawberry in June).

BUYING ORGANIC

Another thing I try to focus on when buying food is organic produce. Organic farming prohibits or severely restricts the use of synthetic pesticides and fertilisers. Its practices also promote sustainability by focusing on soil health, water conservation and reducing polluting techniques.

Organic standards prohibit the use of genetically modified organisms (or GMOs), and you will find that many organic farmers are smaller, family-run operations, so purchasing organic produce can also help support these farmers and encourage more sustainable agricultural practices.

SIMPLE, SUSTAINABLE SWAPS

When I first set out to reduce the plastic in my kitchen, it was purely from a practical point of view, but I soon found that it benefitted me in other ways. The lack of clutter scratches my brain in such a positive way and makes day-to-day living so much smoother. Plus, a sustainable kitchen looks much more appealing to me too.

The following pages contain some really simple swaps we can all make to kick-start our plastic-free practices.

Let's start with the kitchen sink. From the 'disposable' sponges that last only a few days and release microplastics with each use, to the countless bottles of washing up liquid, dish brushes and microfibre cloths, the plastic options in this part of the kitchen seem endless. If you find yourself in this predicament – fear not! I've got some great options for you to avoid the weekly habit of restocking these things.

Washing up liquid

Did you know that most of the 'supermarket' washing up liquids that you can buy contain chemicals that are harmful to the environment? If you have a local refill shop, use a glass bottle to restock the washing up liquid from an eco-friendly brand. (See Resources on page 252 for brand recommendations and more information about sustainable brands.)

If you are unable to refill at your local refill shop, another option is to purchase a solid dishwashing soap bar. I have one of these by my sink for smaller washes and they are great! You just activate them by dipping into your dishwater and rubbing with your brush or sponge.

Another recent discovery that I have made is refillable foaming washing up spray and powders. There are eco-friendly companies online who provide starter packs which contain refillable bottles. You refill these with a small bottle of concentrate and top up with water. This avoids the carbon footprint of shipping water (which tends to make up over 95% of the cleaning products that we use around our home) and also avoids the single-use plastic bottles too.

Dish brushes
Replace your plastic dish brushes with wooden alternatives that have Tampico bristles. You can find many brands that have replacement heads and all you need to do is swap the head over when it is ready to be replaced. You may be sceptical about how durable a dish brush with plant-based bristles can be, but I assure you that they last for months! I also like to use a stumpy little pot brush that is the sturdiest of utensils and really helps with those tough washing up tasks. I can't actually remember the last time I replaced my pot brush; it's been years and still going strong! Every couple of weeks I give my brushes a little clean and they are good to be used again and again.

MY PLASTIC-FREE . . .

Cleaning your natural brushes

You can help your natural dish brushes last longer by cleaning the bristles with bicarbonate of soda using the following steps:

1. Shake off your brushes to remove any loose debris or dust.
2. Add 2–3 tablespoons of bicarbonate of soda and a small amount of dish soap to hot water in a bowl or jug.
3. Submerge the bristles into the solution and let them soak for 10–15 minutes to allow the bicarbonate of soda and soap to break down the dirt and neutralise odours.
4. If necessary, gently scrub the bristles.
5. After soaking and scrubbing, rinse the bristles thoroughly with lukewarm water to remove all traces of the bicarbonate of soda.
6. Shake off any excess water and allow the brush to air-dry completely by hanging it up or laying on a flat surface with the bristles facing down – this will avoid any moisture being retained in the brush.

Kitchen sponges

In years gone by, I used to buy large packets of disposable sponges; the ones that had a yellow sponge on one side and a scouring green side on the other without really thinking how bad they might be for the environment. They were very cheap and designed to be thrown away and replaced after only a few uses, which is exactly what I did, replacing them with a new packet every time I did the shopping. What I didn't realise was that the sponges themselves release microplastics every time they are used, they aren't recyclable and, instead, they go straight to landfill, where they take up to 500 years (!) to degrade.

I made a simple change to plant-based cellulose sponges, and I wonder why I hadn't learned about them sooner. They are a more expensive option, but they are much more durable than their plastic-based counterpart and last for months. The best thing about them is that you can just cut them up and compost them after you have finished using them.

Pot scrapers

To avoid food getting stuck in your washing up bristles and sponges, I recommend a wooden pot scraper to scrape off any food on your plates and pans before washing them. This completely avoids the food getting into your bristles and keeps your utensils nice and clean. I also use this handy tool as a dough scraper when making bread and pizza dough.

Bicarbonate of soda

A slightly different tool to have by the kitchen sink, but an extremely useful one. If you live in a 'hard water' area, adding a tablespoon of bicarbonate of soda to your dishwater will make it softer and easier to clean dishes.

If you have greasy dishes, it's also great to sprinkle on your dishes and soak up the residue before washing. It cuts through grease with ease and is usefully abrasive without causing scratches.

For tough, stuck-on food you can make a paste of bicarbonate of soda and water. Apply it to your pans, let it sit for 20 minutes and then use your pot scraper or scrubbing brush to clean it without causing any scratches.

Spoons and utensils

Opt for wooden or stainless-steel utensils over plastic – they will last longer and won't run the risk of melting while you're cooking. (I'm speaking from experience!)

I love to pick up natural cooking utensils from independent shops. My favourite wood is olive wood as the grain is so beautiful. Olive wood is usually used from olive trees that have stopped bearing fruit which I like, as I feel it gives the trees another life. I give my utensils a good clean and treatment every so often to keep them at their best. By doing this you will find that your natural utensils will last for years.

Caring for your utensils

Wooden utensils need a little attention to keep them at their best, but will last for years if looked after properly. I inherited some of my nana's utensils and they are some of my favourite and most prized possessions; by caring for them this way they are still in good use today.

> 1 lemon, halved
> 2 tablespoons salt
> Dish soap
> Wood balm or walnut oil

1. Dip the lemon half into the salt and use this as an abrasive scrubber on all of your wood utensils and chopping boards. The lemon helps to naturally disinfect and smells amazing, and the salt is naturally abrasive too.

2. After you have thoroughly scrubbed your utensils and boards, rinse them under warm water with a little dish soap.

3. Allow your utensils to thoroughly dry.

4. Once dried, use a cloth to massage either the wood balm or walnut oil* into your utensils.

*If you are using an oil to condition your utensils, it's important to use one that does not go rancid, which is why I suggest using walnut oil. If you have an allergy to one oil, then wood balm would be a good option. (See Resources on page 252 for wood balm recommendations.)

Chopping boards

Plastic chopping boards add microplastics to food every time they are used, so I recommend making the switch to wood or glass. When preparing items like fish or meat, I prefer to use a chopping board that is non-porous. I actually keep a glass chopping board out in the kitchen as it also makes an excellent heat-resistant surface for pans while I'm cooking, and wipes clean really easily.

INVESTING IN REUSABLES

Reusables have helped my family and I to avoid buying so many disposable items such as coffee cups, water bottles and food packaging. I have also found that they save us money too, as we tend to buy less and less when out and about. It's important to mention here that not all reusables are of the same quality, and I did in fact buy some items that were cheap and they broke before I realised that investing that little bit more in reusables would help save us money and waste in the long run. My advice is to take your time researching before buying. If you need a little help making your decisions, here is what I have learned along the way.

Use what you have

While you are taking the time to select the reusables that you would like to invest in, my advice would be to use any items you already have in your home until you've made the decision. For example, to transport my smoothies and water to and from work, I started by using a glass jar I had in the cupboard. To this day, this has worked well for me, so I have never seen the need to replace it. Have a look around your house to see what could be of use, and to see if you can give a supposedly 'disposable' item a second life. Items such as fruit containers can be used as seedling pots for starting your seeds off at spring time, jars can be used to store and transport food, and bags can be used as bin liners – even toilet rolls can be reused for crafts. I share lots of ideas throughout the book, and I hope it may bring some inspiration to see what can be reused before it goes to the recycling.

Reusable bags
The key to switching to reusable bags is remembering to take them out with you when shopping! I store some of my favourite cotton and linen bags in two places – on a coat rack next to the front door, and in the car. This makes them easily accessible and gives me a better chance of remembering them.

Cloth napkins
We have never personally been a household that has ever used disposable kitchen towels. I think when I was a child, they were just an expense that my mum didn't buy into and as an adult I always felt the same. We have a large basket of tea towels that we use as an alternative to oven mitts, dish towels, and for covering food and general cleaning/mopping up messes. We just use them and wash them as we go, so I have quite the mismatched collection.

If you are a fan of kitchen towels and would like to look for a sustainable alternative, you can buy rolls of reusable towels, usually made of organic cotton, that you just pop onto your kitchen towel handle. When you finish a roll, you can wash them in the washing machine and then roll them up again for their next use.

Reusable cleaning cloths
I'm always surprised to see disposable wipes used in adverts as they are just so wasteful but advertised as a necessity. In our house we use cloths for cleaning and polishing, and I also use reusable Swedish dishcloths, which are like a cross between a dishcloth and a sponge. I have them in a number of colours and use a colour code system for the different jobs in our house.

The Swedish dishcloths are made from plant-based cellulose fibres. They are great for mopping up spills, cleaning and all your general household tasks, and again, you can pop them in the washing machine after you've finished using them, ready to be used again.

Reusable straws
Luckily, plastic straws have been banned in many countries, which is great because they were just such an unnecessary plastic waste. There are so many options for reusable straws that you can buy in place of the paper straw, which cater to different needs.

- **Silicone** – Made from food-grade silicone, these are a flexible alternative to a lot of reusable straws that are available. Many come in a little tin for easy transportation.

- **Stainless steel** – Stainless-steel straws are built to last a lifetime and are another great option to use both at home and when travelling. You can usually buy them in different sizes so they can be used for thicker drinks like smoothies too.

- **Glass** – Glass straws are the newest renewable option on the market and are gaining popularity for the more 'aesthetic' drink, like cocktails. Again, they come in different widths so you can choose the size that works for you, and they look great with your glassware. They do not travel well though so more of an 'at home' straw for adult drinks only.

- **Bamboo** – Bamboo is also a great option. I personally love it as it is a super-fast-growing grass that is renewable as it grows back when it has been cut down, unlike trees.

... KITCHEN

Reusable coffee cups
I simply couldn't survive without our reusable coffee cups, and I take them everywhere. They've definitely helped me to save money over the years too. There are so many options out there and I do have my favourites. First and foremost is my stainless-steel, insulated tumbler, which fits perfectly into the cup holder of my car, keeps my hot drink super-hot but also keeps a cold drink cold, and it's simple and easy to clean. Another favourite is an insulated travel flask that is sealed so that I can throw it into my bag when I'm out and about and I can trust it not to leak.

Insulated water bottles
There are so many refill stations these days that you really don't need to buy bottled water, which is a great thing because bottled water has been found to contain microplastics. There are plenty of reusable bottles on the market today that keep your cold water cold for at least 24 hours and your hot water hot for 12 hours. And I recently discovered they keep Prosecco cold and fizzy!

Containers out and about
Having reusable food containers is so handy to avoid single-use plastics like lunch bags and clingfilm, and is also a much safer way to transport your food, no matter where you are going.
I tend to use stainless steel as it is a highly durable option. It is resistant to scratches and rust (in fact, I have stainless-steel food containers that have lasted me for years and still look exactly the same as when I first bought them), it doesn't leach chemicals into your food, which makes it a safer choice for storage, and it is also really easy to clean and can be put in the dishwasher.

Containers at home

For food storage in the fridge, I tend to opt for glass reusables. Again, glass is non-porous and doesn't leach chemicals or react with acidic or alkaline foods when storing. This means the food and quality remain unchanged when using them. Glass is also made from natural materials (sand, soda, limestone) and is fully recyclable – in fact, it can be recycled indefinitely without loss of quality. I have a collection of stackable glass containers with bamboo lids, which I love because I like to see the food inside and they stack so well, keeping everything neat and tidy in the fridge. I also reuse glass jars for smaller items; I have two favourite types for this purpose, a trusty jam jar and a large pickle jar.

For the freezer, I tend to favour silicone as a reusable alternative to freezer bags. Silicone is extremely durable, resistant to heat and cold, and it also has a long lifespan, reducing the need for frequent replacements. I also freeze smaller amounts of food in glass jars. Additionally, I like to use wax wraps to cover bowls or wrap small items to keep them fresh. Wax wraps are pieces of cloth that have been covered with wax and can be used to replace cling film. I find they keep cut fruits and vegetables nice and fresh, too.

MY PLASTIC-FREE ...

Freezing food using jars

Freezing food in jars is one of the most economical and convenient ways to preserve your food, but it also requires a little preparation to avoid issues like jar breakage:

1. Make sure that any food that you freeze in jars is cool before freezing as rapid temperature changes can cause glass to crack.
2. Leave at least 3cm of headspace at the top of the jar to allow for expansion as food will naturally expand as it freezes.
3. Do not fill past the shoulders of the jar and add the lids loosely until fully frozen to reduce pressure on the glass when freezing. Once the food is fully frozen, you can then tighten the lid.
4. When you are ready to defrost the contents of a jar, defrost slowly, either in the fridge or at room temperature. Avoid any rapid changes in temperature, such as placing the jar in hot water or microwaving it.

Reusable silicone baking mats

It wasn't until a few years ago that I learned you couldn't recycle parchment paper, and so I now use a reusable silicone baking mat instead, which I've actually found to be far more efficient. They are very simple to use and you just place them in your baking tray as you would do your parchment paper.

RECIPES

Milk

As a family of four, we do what we can to reduce our reliance on animal products as much as possible; I'm sure you have seen all the statistics about the negative effects eating meat has on the planet, and the impact of the dairy industry, too.

At the stage of our lives we are in, this means that I have the local farmer deliver whole milk for my children, which comes in glass bottles, and my partner and I have plant-based milk. I enjoy making our plant-based milk from scratch, as I found that shop-bought plant milks are not only expensive with difficult-to-recycle packaging, but they also contain additional preservatives, emulsifiers and stabilisers.

Here are three of my favourite recipes, and also the quickest tip you will ever learn for making your own plant-based milk. If you don't have time to make your own, you could see if your area has a local milkman, or if there is a larger company that can deliver dairy or plant-based milks in glass bottles to your door.

Oat milk

Oat milk was the first plant-based milk I ever made. It is the cheapest and probably most popular option, but I have to admit, my initial attempts at making it did not turn out well! It does have a tendency to go a little slimy if you don't follow the right instructions. Luckily for you I have perfected them and now you can too.

> *Makes about 1 litre*
> 110g organic jumbo oats
> A pinch of salt
> 1 teaspoon honey/maple syrup or 1 date, to taste
> ½ teaspoon vanilla extract, to taste
> 1 litre ice-cold water

1. Add the oats and salt to a blender. If you like a sweeter milk, this is also the time to add the honey or maple syrup and vanilla extract. Top with the ice-cold water and blend on high speed for a maximum of 30 seconds.

2. Strain your mixture through a fine-mesh sieve or nut milk bag. It is important that you don't squash the bag too much at this point and allow the milk to gently move through the mesh. This could take up to an hour for the milk to pass through, so if you have the space in the fridge, you could pop it there while waiting.

3. Transfer to a clean, sterile, reusable bottle for storage.

This milk will keep for up to 4 days in the fridge. I use organic oats in this recipe after discovering that oats are often treated with pesticides just before they are harvested and have been found to absorb large amounts of chemicals. You will need to give the milk a little shake before you use it as the natural ingredients will naturally separate without an emulsifier.

    I enjoy this milk in smoothies and on my morning cereal. You will find some leftover pulp from making this recipe, but don't throw it away. Turn to page 40 to learn how to make use of it.

MY PLASTIC-FREE ...

Almond milk

Almond milk has a slightly nutty flavour and is great for adding to your morning coffee. Again, you can sweeten this milk if you prefer, and I have added options to the recipe for you. I also find that almond milk tastes good with a pinch of cinnamon too.

> ***Makes about 1 litre***
> 120g almonds
> 1 teaspoon salt
> 1 litre cold water
> 1 teaspoon honey/maple syrup or 1 date, to taste
> ½ teaspoon vanilla extract (optional), to taste
> A pinch of cinnamon (optional), to taste

1. Place the almonds in a large jar, fill with water, add the salt and soak overnight. If you are rushed for time, you could soak the almonds quickly by adding them to a pan and covering with water, bringing them to the boil and leaving to cool for 1 hour.

2. Drain the almonds and add them to a blender with the cold water. Add the honey/maple syrup or date, according to your taste. Add the vanilla extract and cinnamon, if using. Blend for 30 seconds on high speed.

3. Strain the mixture through a fine-mesh sieve or nut milk bag.

4. Transfer to a clean, sterile, reusable bottle for storage.

This freezes well into ice cubes, so if you can't use it within 4 days, make sure you freeze it. When defrosting, there will be some natural separation, so you will need to shake it to combine it again.

Cashew milk

I would say that this is the creamiest plant-based milk that you will make; it's a great addition to coffee or even to drink on its own. It is a little more expensive than the other options, but will blend completely with no pulp left over.

> *Makes about 1 litre*
> 120g cashews
> 1 teaspoon salt
> 1 litre cold water
> 1 teaspoon honey/maple syrup or 1 date, to taste
> ½ teaspoon vanilla extract (optional), to taste
> A pinch of cinnamon (optional), to taste

1 Place the cashews in a large jar, fill with water, add the salt and soak overnight. If you are rushed for time, you could soak the cashews quickly by adding them to a pan and covering with water, bringing them to the boil and leaving to cool for 1 hour.

2 Drain and rinse the cashews thoroughly until the water runs clear.

3 Add the cashews to a blender with 250ml of the cold water and blend to a paste. Add the remaining 750ml water along with the honey/maple syrup or date, according to your taste. Add the vanilla extract and cinnamon, if using.

4 Blend for 30 seconds on high speed.

5 Transfer to a clean, sterile, reusable bottle for storage.

This milk will keep for up to 4 days in the fridge. As it is a little more expensive than the other milks, I tend to save it for adding to my coffee as a creamer or for those desserts that need a little extra creaminess. This also freezes well into ice cubes, so if you find you have some left over and can't use it within 4 days, make sure you freeze it. When defrosting, there will be some natural separation, so you will need to shake it to combine again.

How to use your leftover pulp: waste-saving energy balls
We love food and hate waste in our home and this is a really simple way to use up the leftover pulp from making plant-based milks.

> *Makes approx. 12 balls*
> Leftover pulp from your plant milk (approx. 150g)
> 85g organic porridge oats
> 40g chocolate chips or cacao nibs
> 1 tablespoon chia seeds
> 60g nut butter
> 60g dates
> 1 tablespoon coconut oil
> 1 tablespoon honey or syrup of your choice

1. Add the leftover pulp, oats, chocolate chips or cacao nibs and chia seeds to a bowl and mix together.
2. Add the nut butter, dates, coconut oil and honey or syrup and blitz to a paste. Add to the bowl and mix well.
3. Roll the mixture into balls approximately 3cm in diameter.
4. Store in an airtight container in the fridge and chomp on one when it takes your fancy.

These will store in the fridge for up to 4 days, if they last that long!

Super-quick nut butter milk

If you didn't know this quick and simple tip to make your own nut milk, it may well change your life! All you need to do is buy a good-quality jar of your favourite nut butter. Make sure that it doesn't contain any additional ingredients; you are looking for a pure nut butter.

> **Makes about 1 litre**
> 2 tablespoons nut butter (I have found that almond, cashew and hazelnut work well)
> 1 litre cold water
> A pinch of salt
> ½ teaspoon honey/maple syrup, to taste
> ½ teaspoon vanilla extract, to taste

1. Add the nut butter to a blender and top with the cold water.
2. Blend for 30–60 seconds on high speed.
3. Transfer to a clean, sterile, reusable bottle for storage.

This milk will store in the fridge for up to 4 days. The first time I made nut milk this way I was honestly astonished, it's so simple and perfect when you find yourself in a pinch and need to make your nut milk extra quickly.

No-waste vegetable stock

I love to make my own vegetable stock as it is a great way to save on food waste, as well as make something that will elevate meals. I just wash and freeze my vegetable scraps as I'm cooking throughout the week and add them to a large container in the freezer.

> *Makes 1 large jar of stock*
> Vegetable scraps*
> 2 tablespoons vegetable oil
> 2 bay leaves
> A pinch of salt

*Vegetable scraps that are good for broth include: onion, garlic, tomatoes, peppers, carrots, celery, leek, herbs and ginger.
*Scraps to avoid include: Brussels sprouts, broccoli and cauliflower, as these can make your stock bitter.

1. As you are cooking throughout the week, make sure you save your vegetable scraps by rinsing them and adding them to a container in the freezer.
2. When you have enough scraps to fill a large pan, brown them off in the oil for a few minutes to get the flavours going.
3. Cover with water, add the bay leaves and salt and simmer for 1–2 hours (you could also use a slow cooker for this).
4. Pass the stock through a sieve and discard the vegetable scraps. It is now ready to use in all of your meals.

You can keep a jar of this in your fridge for up to 4 days; alternatively I also like to freeze it into ice cubes so that I have it to hand. If storing in the freezer it will last for up to 3 months.

I try to add a mix of vegetable scraps as it adds depth of flavour, and I also try to add more of the 'sweeter' vegetable scraps to stop the broth from tasting bitter. Remember that you can also compost the remaining scraps when you have made your stock, and because they have gone through this process, they will also break down faster in your compost.

Snacks to save time, waste and money

I have a number of really simple recipes that I stick to in the kitchen. They are what I call my 'foundation recipes' as I can make them quickly by heart, then add different extra ingredients each time, depending on how I feel and how nutritious I want them to be. In each case I have given the basic recipe first, followed by any optional 'extras' so that you can pick and choose.

Easy biscuits

> *Makes 20 biscuits*
> 200g unsalted softened butter or margarine
> 200g caster sugar
> 1 large free-range egg
> 400g plain flour*
> 50g extra ingredients (optional)

*370g if using any extra ingredients (see opposite)

1. Preheat the oven to 200°C/180°C fan/gas mark 6.
2. Beat the butter or margarine until soft and creamy, then add the sugar and egg and beat together.
3. Add the flour to make a dough and any extra ingredients, if using.
4. Roll the mixture out on a floured surface until about 1cm thick.
5. Cut the dough into strips or cut out shapes. Reroll any spare dough and repeat the process until it has all been used.

6 Place on a reusable baking liner and bake for 8–10 minutes. The biscuits will be ready when the edges start to turn golden.

7 Remove from the oven and allow to cool before eating.

> *Optional extras*
> Add a bit of variety by substituting 30g of the flour with 30g of any of the options below:
> Chocolate chips
> Cacao nibs
> Raisins
> Dried cranberries
> Mixed peel and ½ teaspoon ground cinnamon
> Lavender buds and 1 tablespoon lemon zest

These biscuits are so versatile. We are an ingredients household and my main focus is to make convenient and delicious food every day that the whole family enjoys and looks forward to. As I make most meals from scratch in our house, I don't feel concerned with our family enjoying the occasional bit of sugar or sweet treat. If you are looking to avoid sugar, I do have some alternative basic recipes that are simple to throw together and easily adaptable coming up.

Easy oat biscuits

I love to make these oat biscuits, and my children love them too! They are crunchy on the outside, soft on the inside and taste delicious. A little tip to make these even easier is to double the mixture and freeze half, then make them straight from the freezer.

> *Makes 10 large biscuits*
> 100g softened butter or margarine
> 80g caster sugar
> 100g organic porridge oats
> 100g wholemeal flour
> 1 tablespoon milk of your choice
> 1 tablespoon honey
> 30g extra ingredients (optional – see opposite)

1. Preheat the oven to 200°C/180°C fan/gas mark 6.
2. Beat the butter or margarine until soft and creamy, then add the sugar and beat together.
3. Add the oats, flour, milk and honey to make a dough and any extra ingredients, if using.
4. Roll the mixture out on a floured surface until about 1cm thick. Cut the dough into strips or cut out shapes; reroll any spare dough and repeat the process until it has all been used.
5. Place on a reusable baking liner and bake for 8–10 minutes. The biscuits will be ready when the edges start to turn brown.
6. Remove from the oven and allow to cool before eating.

Optional extras

This recipe happily takes about 30g of any (or a mixture) of these ingredients, so you can play around with it!

Chocolate chips

Cacao nibs

Raisins

Dried cranberries

Mixed peel and ½ teaspoon ground cinnamon

Chia seeds

Mixed seeds

Easy breakfast bars

These are super-nutritious little bars, and I find that they store and travel well in lunch boxes too. Breakfast bars are so delicious and are packed with protein, fibre and healthy fats to keep you satisfied for hours. Again, feel free to make the basic version of these bars (they are delicious), but if you'd like to change it up and add any extras, I have included some options.

> **Makes 6 bars (or 12 mini bars)**
> 60g mixed chopped nuts
> 90g porridge oats
> 70g dates
> 70g prunes
> 1 tablespoon honey or syrup of your choice
> 1½ tablespoons nut butter
> 20g extra ingredients (optional)

1 Toast the nuts and porridge oats in a dry pan until lightly golden then allow to cool.

2 Place the dates, prunes, honey or syrup and nut butter in a blender or food processor and blitz to a paste. You can add more water if it needs help reaching a paste-like consistency.

3 Place the nuts, oats and date mixture into a large bowl and mix to combine. Add any extra ingredients, if using.

4 Press the mixture into a greased baking tray or onto a reusable baking liner and pop it in the fridge to chill.

5 When cooled and set (this should only take about 30 minutes), chop into 6 medium-sized bars or 12 mini bars.

Optional extras
The recipe happily takes about 20g of any (or a mixture) of these ingredients, so switch it up and see what you prefer.
Chocolate chips
Cacao nibs
Raisins
Dried cranberries
Mixed peel and ½ teaspoon ground cinnamon
Chia seeds
Mixed seeds

Because this is a no-bake recipe, I like to toast the oats and nuts first to give the bars that little added flavour. You could also make these into smaller bars for children, if you like. I often make double the mixture and – you guessed it! – freeze the bars that I know we won't eat so that we have them for the second half of the week. These bars will last in an airtight container (I keep them in the fridge) for up to 5 days, or 3 months in the freezer.

Quick granola

This is a delicious and healthy granola recipe that you can make easily and store in a large jar ready to dip into at any time of the day. I personally enjoy eating it heaped on top of yoghurt or have a tasty handful as a snack. This is a really yummy granola that is rich and nutritious and can easily be adapted to your taste by adding optional extras.

> **Fills 1 large Kilner jar**
> 40g coconut oil
> 80g honey or syrup
> 1 teaspoon vanilla extract
> 400g jumbo oats
> 150g mixed nuts
> 50g mixed seeds
> ½ teaspoon ground cinnamon
> A small pinch of salt
> 150g extra ingredients (optional)

1. Preheat the oven to 180°C/160°C fan/gas mark 4 and line a baking sheet with a reusable silicone mat.

2. In a large pan over a low heat, warm the coconut oil, honey or syrup and vanilla extract, stirring gently until the oil has melted.

3. Add the oats, nuts, seeds, cinnamon and salt to the pan and thoroughly combine.

4 Spread the mixture out evenly on a baking tray. Depending on the size of your trays, you may need two, as it is best baked in even but quite thin layers.

5 Bake for 25 minutes, stirring halfway through the cooking time to make sure the granola is toasted evenly.

6 Remove from the oven – your granola should be toasted and smell delicious – and allow to cool.

7 Once the granola has cooled, you can add any extra ingredients, if you like.

8 Thoroughly combine and store in an airtight container.

> *Optional extras*
> The recipe happily takes about 150g of any (or a mixture) of these ingredients, so you can tailor it to make your very own granola perfection!
> Chocolate chips
> Cacao nibs
> Raisins
> Dried cranberries
> Mixed peel
> Chia seeds
> Flax seeds

This will last for up to 2 weeks in an airtight container; I like to store mine in a Kilner jar and keep it on the shelf. Be prepared though, as soon as your family see the full jar, it will be emptied very quickly, and you might find yourself needing to make double the quantity!

MY PLASTIC-FREE ...

COOKING FROM SCRATCH QUICKLY

I love cooking with fresh produce, but sometimes I find that I just don't have time to start my dishes from scratch. Here are a few waste- and time-saving hacks that I love to do for the days that I just need to get some nutritious food on our dinner plates as fast as possible – without compromising on the taste.

Key ingredients prep

Two things I always try to have available in the freezer to start my meals off super-quickly are garlic and onions. They are the start to almost every meal that I make but can take time (that I sometimes don't have) to prepare.

So instead, what I do when I have visited the grocers is I chop all my onions and mince all my garlic at once and freeze it. I promise you, your future self will thank you for it, and it will turn meal prep into a doddle.

Garlic

Taking the skin off garlic can be tedious, so here is a tip for doing it as fast as possible. Get your bulb and cut off the end at the root. This will give you individual cloves without having to rip them off one by one. Now you want to add the whole bulb to a large jar and shake vigorously. You will notice that you've shaken the skins off! Transfer your (now peeled) cloves to a blender and blend until smooth. You will now have potent minced garlic ready for the freezer. Add equal parts oil to the mixture and then pour into an ice-cube tray. You will now have the perfect-sized portions of garlic ready to start any meal.

Onions

This is such a simple thing to do, you just need to peel and chop those onions up and put them into a freezer bag. Then you can take out as much or as little as you like when making your meals. A handy tip is to freeze the onions for a couple of hours and then give them a good shake so that they don't clump together. Or freeze flat onto a reusable baking sheet and then put them into your freezer bag afterwards. I typically use either compostable freezer bags or silicone freezer bags as an alternative to plastic and they work perfectly.

Flavour bombs

As we make the majority of our food from scratch, I like to make these 'flavour bombs' that are a great starter for your dishes and pack a huge amount of flavour. When I make these, I keep a little in the fridge for meals during the week, then freeze the rest in cubes if I haven't used them all. They will last in the freezer for up to 3 months.

These are really simple to make, pack a huge punch, and can be made with lots of different flavour combinations to suit your dishes. It's as simple as grabbing a handful of your favourite herbs and flavourful veggies, adding a couple of glugs of oil and blitzing to a paste. I have three basic starters, a garlic and chilli base, a garlic, tomato and olive base, and a garlic and herb base. (It is all about the garlic in this house!)

Garlic and chilli

I make a jar of this to keep in the fridge and use it to start soups, chillies, curries, as a marinade for roasted vegetables, and anything else that you can think of! It is so flavourful and delicious and really versatile to use, I learned that this is commonly referred to as a sofrito in Spanish and Mexican households and is a favourite way of starting dishes.

> 2 chilli peppers (I use jalapeño peppers)
> 1 small sweet pepper
> 1 small onion
> 1 small handful of fresh coriander
> 8 garlic cloves
> 2 tablespoons olive oil

1. Roughly chop all the ingredients.
2. Place in a blender or food processor along with the olive oil and blitz to a rough paste.

You can store this mixture in a clean jar for up to 1 week in the fridge, or transfer to ice-cube trays and freeze for up to 3 months. This is a simple base recipe that you can adapt to your household. The aim is to have something in the fridge that you find useful to start off your dishes and make flavoursome food without needing extra packets, marinades and sauces, which are often full of preservatives and packaging that we just don't need in our homes.

. . . KITCHEN

Garlic, tomato and olive

This is a Mediterranean version of a flavour bomb that I use to start pasta dishes, vegetable dishes, soups and sauces. It is also delicious as a sandwich spread before you add your fillings, or spread on sourdough bread then topped with eggs or avocado – yum!

> 8 sun-dried tomatoes
> 4 garlic cloves
> 125g pitted black olives
> 2 tablespoons olive oil
> A handful of basil leaves

1. Place all the ingredients in a blender or food processor and blitz to a rough paste.
2. Use as a base to start your meals or drizzle on top of cooked dishes to add extra flavour.

You can store this mixture in a clean jar for up to 1 week in the fridge, or transfer to ice-cube trays and freeze for up to 3 months.

Garlic and herb

These flavour bombs are a great way to use up any of those woody, flavourful herbs that you may have extra of. I freeze these into cubes for up to 3 months and add them to a frying pan or ovenproof dish when starting so many dishes. It is my opinion that garlic and herbs belong in nearly all dishes!

> 1 bulb of garlic
> 2 handfuls of woody herbs, such as rosemary and thyme
> 125ml olive oil

1. Place the garlic, herbs and a few glugs of the olive oil in a blender or food processor and blitz to form a paste.
2. Top up with the rest of the olive oil and stir through.

Again, this mixture can be stored in a clean jar for 1 week or frozen into cubes. My favourite way to use these garlic and herb bombs is simply roasting vegetables in the oven – delicious.

FOOD STORAGE AND REDUCING WASTE

The right food storage can help to extend the shelf life of your food, saving you time and money, and protecting the environment too. I tend to follow these rules:

Clean your vegetables

I always wash and dry vegetables before storing them. As well as preventing them from going 'bad' this also helps to reduce the number of pesticides on your veggies. It also means that you can just grab and go from the fridge without the additional step of washing when you prepare a meal.

1 Fill a sink with cold water, add 2–3 tablespoons of bicarbonate of soda and give the vegetables a little bath for up to 10 minutes.
2 Using a natural vegetable brush, gently scrub the surfaces to help remove stubborn dirt.
3 Add your veggies to a colander and rinse thoroughly.
4 Pat your veggies dry with a soft towel and allow them to air-dry; they can now be stored.

For leafy greens I tend to soak for only a minute as they are more delicate, then I use a salad spinner to dry them. The only items that I don't wash like this are cauliflower, broccoli, mushrooms and soft fruits as I find that it actually reduces their shelf life. I wash these when we want to eat them and share tips on how I store them below.

... KITCHEN

Storing fruits and vegetables

Leafy greens – spinach, kale, lettuce

How to store: Line a large container with a damp, cotton towel. Place the leaves on top and place in the fridge. It might sound a little counterintuitive to use a damp towel, but I promise you that they will keep your greens crisp and fresh for so much longer

Freezer tips: Did you know that you can freeze spinach leaves just as they are? You can just grab a handful to add to your smoothies, curries or soups when you need to! I also freeze them into cubes by blending 4 cups of whatever greens I have to hand (kale, spinach or even lettuce) with one cup of water. Add them to an ice-cube tray and store for whenever you need them.

Root vegetables – carrots, radishes, beetroot, parsnips

How to store: Remove the leafy tops and store in a container in the fridge.

Freezer tips: To prep and freeze, peel and slice them then blanch in boiling water for 2–3 minutes. The carrots will be vibrant in colour and still crisp. Transfer the carrots to a bowl of ice-cold water for 2–3 minutes to stop the cooking process. Place on a baking sheet to freeze; when frozen you can then transfer to a freezer bag.

Potatoes and sweet potatoes

How to store: In a cool, dark and well-ventilated place. Refrigerating can affect the structure and taste of potatoes.

Freezer tips: Potatoes hold a lot of water so shouldn't be frozen raw. They freeze well when roasted or even mashed. You can also blanch your potatoes by cooking them in boiling water for 3–5 minutes, then immediately transferring to iced water. Drain and pat dry and transfer to a reusable baking sheet to freeze, before transferring to an appropriate container.

Onions and garlic

How to store: In a cool, dark and well-ventilated place. Refrigerating them isn't recommended as it can affect the texture and flavour. Keep onions separate from potatoes, as they can cause each other to spoil too quickly.

Freezer tips: See pages 53–54.

Broccoli and cauliflower

How to store: In the fridge and only washed just before they are eaten as the water can cause spoilage.

Freezer tips: Broccoli and cauliflower need a little more preparation before freezing. Cut them into florets and blanch them in boiling water for 2–3 minutes.

The florets should become tender, crisp and vibrant in colour but not fully cooked. After this you will need to transfer them to a bowl of ice-cold water for 2–3 minutes. This will help to preserve the texture and colour of the florets. Transfer to a baking sheet and pop in the freezer. When frozen, you can transfer the florets to a freezer bag.

Tomatoes

How to store: Tomatoes should be stored at room temperature until they are ripe (and this is also the best time to eat them). But if you find you have more than you can eat, you can store them in the fridge to extend their freshness; just make sure you take them out and bring to room temperature before eating to maximise their flavour.

Freezer tips: I freeze cherry tomatoes whole or chop up larger tomatoes and pop them into a container. They are fantastic to pop onto a baking sheet and roast for tasty pasta sauces. You can also make a pasta sauce using tomatoes when they are fresh and freeze the sauce once you have made it. I like to freeze into portion-sized blocks so that they are easy to use.

Peppers, cucumbers and courgettes

How to store: These are best stored in the fridge to prolong their life. I have a cotton 'salad bag' that I use to store peppers and cucumbers.

Freezer tips: Slice or chop up peppers and transfer to a freezer bag.

... KITCHEN

Mushrooms

How to store: In a paper bag in the fridge. I avoid washing mushrooms before I store them as this leads to them becoming slimy. It's best to brush or wipe them just before use.

Freezer tips: You can freeze sliced or whole mushrooms but freezing will change their consistency. Frozen mushrooms are best added to dishes like stews, casseroles, pasta bakes or stir fries.

Asparagus

How to store: Did you know that you can store your asparagus like a little bunch of flowers? Just trim the ends and place them, upright, in a jar with 3cm of water at the bottom. Change the water every few days.

Freezer tips: Asparagus also need a little help before freezing to avoid them going mushy. Add to boiling water and cook for 2–3 minutes until they are bright green and crisp but tender. Transfer to an ice bath to stop the cooking process, drain and pat dry with a cloth to remove excess water. Freeze on a reusable baking sheet before transferring to an appropriate container.

Herbs

How to store: My favourite way to keep herbs fresh is to grow them in the garden and take what I need when I need it. But if you don't have the time or space to grow herbs you can also store these in the same way as asparagus to keep them fresh. Just add them, upright, to a jar with 5cm of water at the bottom. Change the water every few days.

Freezer tips: Chop your leftover herbs up and mix in oil to freeze. See page 54.

Bananas

How to store: Separately from every other fruit or vegetable. Why? Because bananas produce a significant amount of ethylene gas, which is a natural plant hormone that accelerates its ripening process and can also ripen the fruit and vegetables around them more quickly. With this in mind, keep them separate from your other produce UNLESS you need to ripen a piece of fruit or vegetable quickly.

Freezer tips: Make sure you peel them, then chop them in half and pop into a freezer bag. Frozen bananas make a great base for ice cream, smoothies and baking.

Berries – strawberries, raspberries, blueberries, cherries

How to store: I have a handy little tip to help berries last longer. Find a clean container, pop a small, dry towel or reusable kitchen towel in the bottom and place your berries on top. The towel will absorb any extra moisture that usually would spoil them. Don't wash your berries until they are ready to be used.

Freezer tips: All you need to do is wash and dry the berries then spread them on a baking sheet to freeze. Once they have frozen, you can transfer to a freezer bag. They are great on top of yoghurts, added to smoothies or baked in desserts.

Citrus fruit

How to store: At room temperature but if not used within a few days they can be stored in the fridge to extend their shelf life.

Freezer tips: Lemons and limes are great citrus fruits to freeze. The zest and juice can be frozen into ice-cube trays.

Melons

How to store: At room temperature; once cut, store in the fridge.

Freezer tips: Cut your melon into slices, chunks or balls and freeze on a reusable baking sheet before transferring to an appropriate container. Frozen melon works well with juices, smoothies and sorbets. It is also lovely eaten frozen on a hot day.

Stone fruits – peaches, plums, nectarines, apricots

How to store: Store at room temperature but if not used within a few days they can be stored in the fridge to extend their shelf life.

Freezer tips: Remove any pits and freeze on a baking sheet before transferring to an appropriate container. Frozen stone fruit also works well in smoothies and even pies and crumbles.

Grapes

How to store: Store in the fridge in a breathable container.

Freezer tips: Wash and dry grapes and freeze whole. These are also great added to smoothies and make a really refreshing snack too.

Apples and pears

How to store: Store at room temperature but if not eaten within a few days they can be stored in the fridge to extend their shelf life.

Freezer tips: Add to a bowl of salt water for a few minutes to prevent browning. Drain and pat dry and transfer to a freezer-safe container. Frozen apples and pears also work well in smoothies, pies and crumbles.

Pineapple, kiwi fruit, mangoes, papaya and avocados

How to store: Store these fruits at room temperature until ripe, then store in the fridge to extend their shelf life.

Freezer tips: Chop then freeze on a baking sheet before transferring to an appropriate container. All of these fruits will work well in juices and smoothies.

What you can freeze to save food waste

I am a HUGE fan of the freezer; by the end of the week my fridge often looks bare, but my freezer is stocked up full of handy food items. Freezing is an excellent way to extend the shelf life of so many foods and reduce food waste. Here, I'm sharing some more foods that freeze well, and those to avoid putting in the freezer.

Dairy

Milk: You can freeze milk into portion-sized cubes ready to use when you need them. Defrost in the fridge or add straight to smoothies.
Cheese: Cheese can also be frozen as it is and defrosted later. It can also be grated and frozen so that you can take out only what you need with no wastage! Soft cheese also freezes well to be used in cooking.
Butter: Butter can be frozen in its original packaging or portion sizes to be used at a later date.

Eggs

Raw eggs can be cracked and lightly beaten before freezing. They can then be frozen in silicone moulds to be defrosted and used at a later date. Cooked eggs also freeze and defrost well, as well as egg-based meals.

Bakery items

Bread: Bread, bagels, tortilla wraps, crumpets etc. can be frozen in bags. If it needs slicing, make sure that you slice it before freezing.

Muffins, cookies and cakes: Freeze in freezer bags to defrost later.

Meat and poultry

Raw meat and poultry can be frozen into smaller portions ready to defrost and cook when needed. Cooked meat and poultry can also be frozen into smaller portions to be used at a later date. Make sure that any meat is defrosted slowly in the fridge and be careful not to reheat more than once.

Seafood

Fish fillets and prawns can be frozen into smaller portions to be defrosted and used when required.

Grains and pasta

It is very easy to freeze cooked rice and quinoa. Cooked pasta can also be frozen, although it is useful to add a little oil to the pasta so that it doesn't all clump together in the freezer bags.

Soups and prepared stews or casseroles

You can freeze soups and stews into smaller portion sizes so that they can be defrosted easily.

Pizza dough

Portion up your pizza dough and freeze into individual portions. You can then just defrost how many you need.

Nuts and seeds

Nuts and seeds can also be frozen to extend their shelf life – they freeze and defrost well.

Tofu

Tofu can be frozen in its original packaging. It actually makes the texture firmer, which I prefer.

Tips for freezing
It's really handy to label your freezer food. I like to write the date and name of food on the container, so it's organised. I usually try to use frozen food up after 3 months and keep a rotation in my freezer.

Food you can regrow

As well as making sure we don't waste food, did you know that you can actually regrow some of your food from the scraps? I love to do this as it feels so satisfying, reduces extra single-use packaging and also helps to save money too. My children love it too and view it as a little experiment. Here are some ways that you can regrow your food from scraps:

Spring onions: These grow back so quickly, all you need to do is place the white root ends in a glass of water, and they will regrow new shoots in just a few days.

Celery: Cut the base off a celery bunch and place it in a shallow dish of water. New stalks and leaves will regrow from the centre.

Romaine lettuce: Similar to celery, the base of a lettuce head can regrow in water, producing new leaves.

Garlic: You can regrow garlic by planting individual cloves in soil. Each clove will grow into a new bulb.

Carrot tops: You can't regrow the carrot root itself, but you can regrow the greens by placing the top part of a carrot in water.

Leeks: Regrow them just like green onions by placing the white root ends in water.

. . . KITCHEN

Ginger: Soak a piece of ginger root in water overnight, then plant it in soil. It will sprout new shoots and roots.

Potatoes: Let the eyes of a potato sprout, then cut the potato into sections and plant them. Each section with an eye will grow into a new potato plant.

Bok choy: Place the base in water, and new leaves will start to grow from the centre.

Sweet potatoes: Place half of a sweet potato in water, and shoots will grow, which can be planted to produce more sweet potatoes.

Some of these methods will grow back only a small amount of the original vegetable, while others will grow whole, new plants. It's fun to do, and if it's something that you haven't tried before, you will be enchanted by the process!

... KITCHEN

73

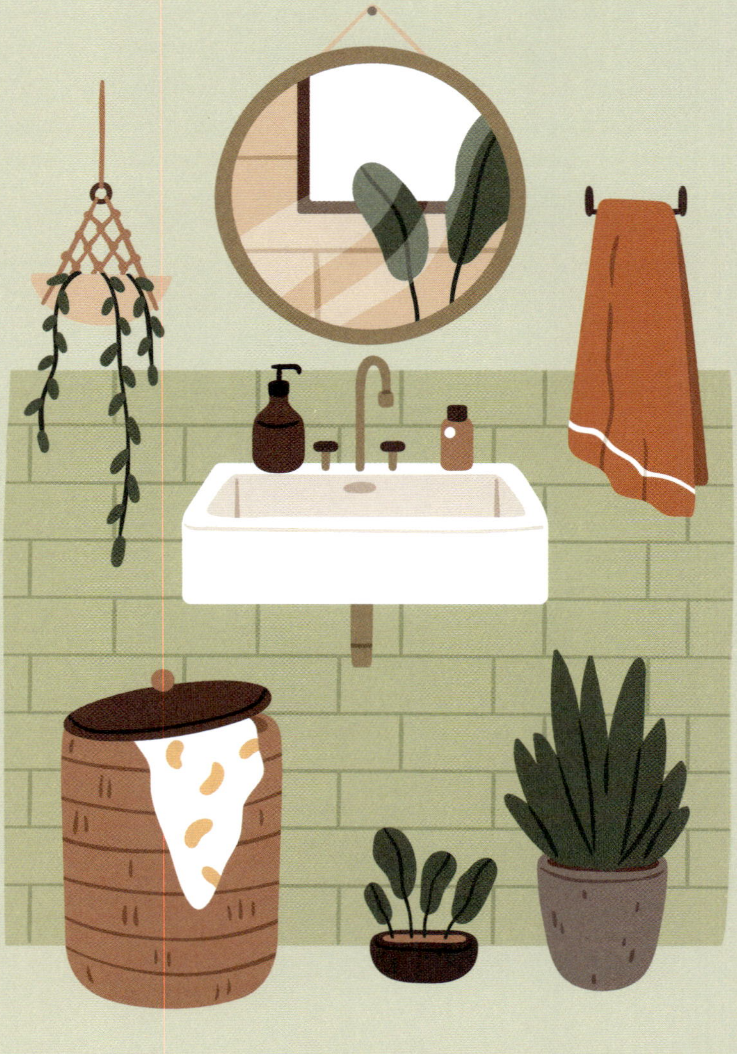

MY PLASTIC-FREE BATHROOM

SIMPLE, SUSTAINABLE SWAPS

It's so easy to get caught in a cycle of feeling that you need a gazillion plastic-packaged products for the bathroom – the number of adverts I see on a daily basis telling me that I need this or that product is testament to this. Over the years I have managed to simplify both my bathroom and skincare routine using both DIY recipes and eco-friendly products. The following pages contain some really simple swaps we can all make to kick-start our plastic-free practices.

Soap

The humble soap bar doesn't get nearly enough credit! It's so versatile, lasts for ages and travels well too. I love to use a natural soap bar and I add it to a sisal soap pouch to help it form a great lather in the shower.

Soap saver pouch

These are small, drawstring bags made from plant-based sisal fibres, which are derived from the leaves of the agave plant. You just place the soap inside and work to help create a lather; they also provide a rough texture that gently exfoliates the skin. They make your soaps last longer too.

... BATHROOM

77

Shower gel

Although I prefer a soap pouch, there are other people in our house who prefer to use a shower gel, and I have a lovely recipe that makes a really luxurious, nourishing, foaming gel. I find it really useful to make my own as I can avoid single-use packaging this way and just store it in a reusable container.

> *Makes 1 x 300ml bottle*
> 200ml liquid Castile soap
> 30ml sweet almond oil
> 20ml vegetable glycerine
> 20 drops of mandarin essential oil
> 20 drops of lavender essential oil
> 10 drops of tea tree essential oil
> 3g Preservative Eco*

*Adding a small amount of natural preservative to this product will keep it shelf stable for up to 3 months. Preservative Eco is an eco-friendly preservative that protects your skincare products from bacterial growth when using water-based ingredients. You can buy this online from natural beauty shops; see Resources on page 252 for recommendations. If you do not have Preservative Eco but would like to try the recipe, I would advise making a smaller quantity, storing it in an airtight container with a pump head and using up within a week.

1. Place all of the ingredients in a jug and mix together until combined.
2. Pour into a clean, reusable plastic bottle with a pump dispenser.
3. Shake before each use.

Konjac sponge

I love to use a konjac sponge to cleanse my face morning and night. It is a great, plant-based alternative to plastic sponges, which are synthetic and shed microplastics. It's gentle enough to use with just a little water or you can also pair it with a cleanser; my favourite is a bi-phase cleanser that I make (see page 98). The konjac sponge is a natural, eco-friendly sponge made from the root of the konjac plant. It appears as hard as a pumice stone when it is dry but softens and becomes almost like jelly when moistened. The sponge provides very gentle exfoliation and is suitable for all skin types, including sensitive skin. It is a natural way to clean pores, remove dirt and oil, and prevent acne.

To use the sponge, you simply need to soak it for a few minutes until it becomes soft and expands. Then gently squeeze out any excess water and use the sponge to cleanse your face in a circular motion.

Be sure to hang the sponge to dry in a well-ventilated area between uses and replace your sponge every 4–6 weeks to ensure it remains effective. You can also periodically disinfect the sponge by boiling it in water for a few minutes.

Loofah

A loofah, also known as luffa, is a key staple in my bathroom to help exfoliate my skin naturally. I actually grow loofahs in my garden (see page 244). They are a relation to the cucumber plant and grow well in a greenhouse.

The loofah is a natural sponge that is formed by the fibrous material inside the mature luffa plant. It is quite rough when dry but softens when wet and the texture of the loofah helps to exfoliate dead skin cells, improving circulation and promoting smoother and healthier skin.

Once you've finished using your loofah it is recommended that you also hang it to dry in a well-ventilated area. You can also disinfect your loofah by boiling it in a pan of water for a few minutes.

Pumice stone

These are a natural alternative for exfoliating and smoothing rough skin areas. They are made from volcanic lava that hardens as it cools, creating a light, porous stone. They remove dead skin cells and help to smooth and soften the skin.

To effectively use your pumice stone, it needs to be wet before being softly rubbed against the skin. Allow the pumice stone to air-dry completely in a well-ventilated area between uses and simply rinse thoroughly after each use. You can also clean it with a brush and soap and periodically boil it for a few minutes to disinfect.

Safety razor

I absolutely love my safety razor, but it's something that I took a little while to try because I was worried about accidentally cutting myself! Happily, I haven't had any accidents yet. Not only that, but it's put a stop to rashes or ingrown hairs, which I always used to struggle with when using disposable razors.

If you haven't used a safety razor before, it may need a little introduction. It is a reusable razor that is made from stainless steel and has a single replaceable blade. There are different styles, but commonly you just unscrew the top of the razor to replace the blade.

After the initial investment, they are really cost-effective because the replacement blades are so inexpensive compared to disposables. Using a safety razor saves waste because you only need to replace the blade rather than the (usually plastic) razor, and they provide a closer and smoother shave than their disposable counterparts.

Using a safety razor:

1 Cleanse your skin in a warm shower or bath to help soften your skin and open up the pores.

2 Apply shaving soap to your skin.

3 Hold the razor at a 30-degree angle to your skin. You do not need to press the razor against your skin as the weight of the razor itself provides sufficient pressure.

4 Use short, controlled strokes in the direction of hair growth (this will reduce the risk of irritation and ingrown hairs). Rinse the blade under warm water after every few strokes to keep it clean.

5 Rinse the area that you've shaved with cold water.

6 Apply a moisturiser after shaving to reduce any irritation and keep your skin hydrated.

After using your safety razor, rinse it under warm water to remove any remaining hair or soap and dry the razor with a towel to prevent rust and prolong the life of the razor. You should store your razor in a dry place and replace the blade after five to seven shaves, or when you feel it tugging on your hair.

Cellulose sponge

Made from wood pulp and plant fibres, cellulose sponges are super-absorbent, durable and completely compostable. They are soft enough to be used as both face and body sponges, and also come in different shapes – my children love the little ducks.

As with all of the natural products that I use in the bathroom, it is important to allow sponges to dry out in between uses. You can also disinfect them periodically by soaking them in a pan of boiling water for a couple of minutes.

Shampoo and conditioner bars

Finding the right shampoo and conditioner has been a matter of trial and error for me, but I am now happy to have discovered some fantastic alternatives.

The simplest way to avoid plastic bottles is to make the swap to shampoo and conditioner bars. They typically contain gentle cleansers as well as natural oils, butters and essential oils. They are also free from sulphates, parabens and synthetic fragrances that you commonly find in supermarket hair products.

Shampoo and conditioner bars typically last two to three times longer than liquid alternatives. They are really easy to take travelling with you; just pop them in a travel tin and off you go.

To use a shampoo bar all you need to do is thoroughly wet your hair and lather the bar by rubbing it between your hands or even directly onto your scalp. Then work the lather through your hair, massaging your scalp like you would with a regular shampoo. Make sure to rinse your hair thoroughly to rid it of any residue.

Conditioner bars are also really simple to use. Ensure that both the bar and your hair are thoroughly wet and massage the bar into your hair, focusing on the ends.

Allow the bars to dry out between uses. Alternatively, if you can't tear yourself away from liquid shampoos and conditioners, but you have a local refill shop, there are a number of refillable options available that form part of a 'closed loop' system. (See Resources on page 252 for further information.)

Leave-in conditioner

Sometimes the simplest things really are the most obvious and this really is one of those recipes. All you need is the conditioner that you use regularly, a spray bottle and water to make a nourishing leave-in conditioner to suit your hair type.

Makes 1 x 100ml bottle
50ml boiling water
40ml conditioner of your choice
1g Preservative Eco*

*When adding water to a simple recipe, it can make that product unstable and will therefore need to be used quickly so bacteria doesn't form in the product. Adding a small amount of Preservative Eco to this product will keep it shelf stable for up to 3 months.

1. Pour the boiling water in a mixing jug and add the conditioner. If your conditioner is in solid form, grate it so that it dissolves into the water.

2. Mix together until completely combined and set aside to cool.

3. When cool add the Preservative Eco to help stabilise and extend the shelf life of your product.

4. Pour into a clean, reusable spray bottle.

If you would prefer to make this without the additional preservative or you do not have it to hand, I would recommend making a smaller quantity and storing it in the fridge. It will be shelf stable for approximately 1 week.

Toothpaste
There are a few varieties of toothpaste that you can use to avoid the traditional plastic tubes:

Toothpaste in glass jars
These are readily available from many eco shops (see page 252). They are available as a paste or powder and taste just like regular toothpaste. You can also find them with or without fluoride.

Toothpaste in aluminium tubes
Aluminium is a natural element and infinitely recyclable, making it a great alternative to plastic. Aluminium toothpaste tubes are also readily available from eco shops.

Toothpaste tablets
Toothpaste tablets are compact, lightweight and easy to pack for your travels. All you need to do to use a toothpaste tablet is pop one into your mount and chew it until it forms a paste, then use it to brush your teeth with.

Toothbrushes
There are so many fantastic alternatives to plastic toothbrushes available, and they last just as long! Although there is not currently an alternative to plastic bristles available, you can buy toothbrushes with bristles that are made from plant-based plastics, which is certainly a step in the right direction. Just make sure to remove the bristles before composting the toothbrush.

My favourite toothbrushes to use are included in the Resources on page 252. I swap between using FSC-certified wooden or bamboo toothbrushes when I am travelling and a sustainable, electric option for day to day. Bamboo is a great sustainable option that is becoming a common replacement for wood and plastic in many products. It is a more sustainable option because unlike wood, bamboo regrows quickly when it is harvested and does not need replanting.

Electric toothbrush

You can now buy bamboo toothbrush heads that are compatible with your existing electric toothbrush, which can then be composted in the same way as bamboo toothbrushes.

I use a stainless-steel, electric toothbrush which can be repaired and recycled (unlike the plastic toothbrushes available and widely used). The heads for the toothbrushes are also made from corn starch and castor oil, and I just send them back to be recycled.

Dental floss

Dental floss is another item in the bathroom that contains plastic that cannot be recycled. But there are some great alternatives that you can use instead. However, you can now buy dental floss that is made from cornflour/cornstarch and is fully compostable.

...BATHROOM

DIY method: oil pulling

Oil pulling is an ancient remedy that's known to improve dental health, freshen breath and whiten teeth. It also helps to reduce the bacteria in your mouth that can lead to tooth decay and gum disease. It's a simple process that's easy to incorporate into your daily oral hygiene practices although may take a little getting used to, and you may want to start with less oil and a shorter duration until you get used to it.

> 1 tablespoon organic coconut oil

1. Swish the coconut oil in your mouth for about 15–20 minutes. (It can feel like a long time at first, but you soon get used to it and can do it while you get on with your morning routine, and the results are worth it!) By pushing the oil through your teeth, you get to every area that the bacteria can hide.
2. Be careful not to swallow the coconut oil as you are swishing.
3. After you have finished oil pulling, the oil can be composted in your compost bin! Be careful not to put it down the drain as this could congeal and cause a blockage.

Tongue scraping

Tongue scraping is a quick and effective way to remove debris, plaque and bacteria from your tongue. It is a method of dental hygiene that has been around for thousands of years. I use a copper tongue scraper.

Using a tongue scraper:

1. Stick out your tongue and place the scraper at the back of your tongue.

2. Move the scraper to the front about 3 times using a light pressure.

3. Swish your mouth out with water after scraping.

Deodorant

Using a natural deodorant instead of aerosols is so much better for you, and the planet will thank you too. Aerosols are full of synthetic ingredients and release harmful chemicals into the air; they can also be harsh on your skin and can cause reactions. I don't think I will ever forget how suffocating that deodorant cloud feels when you spray it in the bathroom, it's another thing I reflect on and think . . . why wasn't I questioning that when I was spraying it directly onto my arm pits?!

Here is a simple DIY deodorant recipe that I like to use. It's so easy to make and you only need a little for it to be effective.

Natural deodorant

> *Makes 1 small jar*
> 30g coconut oil
> 30g bicarbonate of soda*
> 30g cornflour/cornstarch
> 5 drops of lavender essential oil (optional)
> 5 drops of mandarin essential oil (optional)
> 5 drops of tea tree essential oil (optional)

*If you have sensitive skin, I recommend replacing the bicarbonate of soda with arrowroot powder. Bicarbonate of soda is an effective ingredient in deodorant and commonly used, but if you find irritation occurs, arrowroot powder can be an effective alternative.

1. Melt the coconut oil gently in a saucepan.

2. In a separate bowl, mix the bicarbonate of soda and cornflour.

3. Add the melted coconut oil and mix together until you reach a creamy consistency.
4. Add the essential oils, if using. I have chosen this blend as they smell lovely but they are also antifungal.
5. Place in a clean, sterile jar and allow to set overnight.

The deodorant will set solid, but very easily melts into skin when you use it. Just rub a small amount under your arms as needed.

Crystal deodorant

Crystal or mineral deodorants are another great alternative to mainstream deodorants. They are made from mineral salts, which have naturally antibacterial properties that help to control body odours and inhibit the growth of odour-causing bacteria rather than masking the odour with fragrances.

To use them, all you need to do is wet the surface of the deodorant and rub the moistened crystal under your arms and allow to dry. They only contain natural ingredients and are hypoallergenic and long-lasting, making them a really cost-effective alternative. The one I favour comes in a cork case too and is completely plastic-free.

Refillable cream deodorant

Cream deodorants contain no synthetic ingredients and are cruelty free and eco-friendly. Most importantly – they work! You can buy reusable cartridges and refill them with replacement capsules which are packaged in compostable, paper packaging.

Reusable sanitary products

Disposable pads and tampons can take up to 500 years to decompose! Thankfully, there are lots of options for reusable sanitary products on the market. As well as being plastic-free, these reusable products are made with fewer chemicals and more natural materials like organic cotton and bamboo. This also makes them kinder to the skin, leading to fewer irritations and allergies.

Menstrual cup

A menstrual cup is flexible and usually made from medical-grade silicone. It can be used for 6–12 hours before it needs emptying, depending on your flow, and is reusable for up to 10 years! It comes in two different sizes with the smallest size recommended for people who are younger or who have not had children, and the largest size for people who are over 30 or who have given birth vaginally.

When I was at university many moons ago my first discovery of a menstrual cup was a little sticker behind the door in the girl's toilet, and it felt like a secret society! I'm so glad to see that they are so much more talked about now and they are even available in supermarkets.

Period underwear

I'm going to make a bold statement now, and I am quite sure that you are going to agree once you've tried these out. If you're a sanitary towel user and have never tried period underwear before, this is your sign that not only will you find it hands down a better alternative, but your period will have never felt comfier.

Period underwear looks like regular underwear, but it's designed with built-in absorbent layers to hold period flow. It's a little like having a very comfortable and discreet menstrual pad inbuilt into your underwear. They come in so many different styles, sizes and also flow. They are really comfortable and can be worn alone or as a backup to other menstrual products (I wear mine with a menstrual cup on heavy days).

To care for your period underwear, you just need to rinse them with cold water after you've worn them and then pop them in a normal wash with the rest of your washing. It's surprisingly simple!

Cloth menstrual pads
Made from soft absorbent fabric such as cotton, bamboo or hemp, cloth pads come in various sizes and absorbencies and are reusable for several years.

They have been found to be more comfortable and breathable compared to disposable pads. To care for your menstrual pads you just need to rinse them with cold water and then pop them in a normal wash with the rest of your washing.

MY PLASTIC-FREE . . .

DIY SKINCARE

Here are a handful of my favourite simple, rejuvenating skincare recipes that make for great alternatives to plastic packaged skincare products on the market today. They incorporate only a handful of really effective ingredients and can all be made from your kitchen table.

With all of these products, I use quite specific measurements. I find it is useful to use a pair of jewellery scales that measure to a degree of 0.01g for this purpose (you can pick them up quite easily second hand). The reason that I use specific measurements is because the quantity of the product is small and the addition of any preservative or essential oil needs to be accurate to avoid any irritation. I also measure liquids in grams as this keeps the overall recipe more accurate.

Making your own skincare products is easier than baking once you get used to it, the ingredients are so amazing for your skin and you will soon find that you only need a handful of products to feel nourished from tip to toe.

Cleansing butter

This is such a luxurious cleansing butter, which can be used in the morning to prepare your skin for the day, and at night to remove every trace of make-up. I like to massage it into my skin for up to a minute and then use a warm, damp cloth, konjac sponge or cellulose face sponge to remove it from my skin.

This creamy cleanser contains castor oil, which has been in the spotlight lately for all of its skin and body healing properties. I find that my skin is too sensitive to use castor oil as a leave-on treatment, but I do like to use it as part of my cleansing routine to impart its properties. Castor oil acts like a magnet to draw out impurities in the skin.

You may find that when you use this cleanser at night, your face feels so nourished that you don't feel the need to add anything more to it! Sometimes I just finish the routine off with a spritz of rose water, and sometimes I'll add a lotion or face oil.

Makes 1 medium-sized jar
48g mango butter or shea butter
25g jojoba oil
25g castor oil
1g vitamin E oil
0.5g essential oils for skin concerns (optional, see below for recommendations)

- If you have oily, acne-prone skin, you will find it beneficial to add tea tree essential oil to your recipe. Tea tree is a powerful antibacterial and antifungal oil that can help to clear black heads without being too harsh.

- If you have inflamed or irritated skin, you will find it beneficial to add lavender essential oil to your recipe, as it helps calm irritation and inflammation while promoting skin healing.

- If you would like to tighten and firm the skin, you will find that adding rosemary essential oil will be beneficial to your skincare routine. Rosemary essential oil's astringent properties help to tighten and firm the skin, while also helping to reduce bacteria and excess oil.

Within these recommendations for essential oils, I have been careful to only select essential oils that work well on the skin. Please keep to the recommended quantity to use them safely and add benefit to your skincare product.

1. Using the double boiler method, warm the mango butter, jojoba oil and castor oil together until melted.
2. Allow the mixture to cool to room temperature then add the vitamin E oil and your chosen essential oils, if using, and stir to combine.
3. Pour into a clean, sterile jar and allow to set for 24 hours.
4. To use this cleanser simply massage a small amount into the skin, then remove with a warm, damp cloth. A konjac sponge or face sponge would also be perfect for this.

Witch hazel cleanser

This cleanser is super-efficient at removing make-up and helps to balance the oil production of your skin, making it great if you are prone to break outs. This type of cleanser is called a bi-phase cleanser, and you will need to shake the bottle before you use it in order to combine the two ingredients.

Makes 1 x 100ml bottle
59g jojoba oil
40g witch hazel hydrosol
1g Preservative Eco*

* When introducing waters to skincare products there is a risk that bacteria will grow in the product, giving it a limited shelf life of only a few days. Using a natural preservative in any formulation that include waters will make the product safe to use and give it a shelf life of 3 months.

1. Combine all of the ingredients together in a jug.
2. Pour into a clean, sterile, reusable bottle – a spray bottle works well for this so that you can spray it directly onto your skin.
3. Shake before use and remove with a damp face towel or sponge.

Face oil

A simple face oil is so wonderful and rejuvenating for my skin. They are quite expensive to buy in the shops, but you can make them easily at home for a fraction of the price. I use jojoba oil in all of the skincare recipes as it mimics the skin's natural sebum; hydrating your skin without clogging your pores. It's also perfect for keeping your skin balanced, whether you're oily, dry or a bit in between.

Rosehip oil is packed with nutrients and is thought to be like a natural retinol, gently smoothing skin without any irritation. It has soothing, anti-inflammatory effects, making it great for calming irritated or sensitive skin. If using rosehip oil in the morning, make sure you follow up with SPF during the day, or use rosehip oil for your nighttime routine and sweet almond oil in the morning.

Makes 1 x 100ml bottle
50g rosehip oil or sweet almond oil
49g jojoba oil
1g vitamin E oil

1. Pour the rosehip oil, jojoba oil and vitamin E oil into a clean, sterile beaker and stir to combine.
2. Transfer to a clean, sterile, reusable bottle with a pipette.

To apply this face oil, I simply add a couple of drops to the palms of my hands and rub them together to warm the oil, then I massage it into my face. I like to complete a little massage routine when using this oil, which I find helps to give my skin a healthy glow.

Lotion balm

This body lotion is a recipe that I have used for years. It is so simple and you will truly be delighted with how nourishing it is. It is perfect even if you have conditions like eczema, and the ingredients are non-comedogenic, which means that it will not block your pores.

> *Makes 1 x medium jar*
> 45g mango butter or shea butter
> 53g jojoba oil
> 2g vitamin E oil

1 Using the double boiler method, gently melt together the mango butter and jojoba oil until it is liquid and combined.

2 Allow the mixture to cool to room temperature then add the vitamin E oil and gently whisk together. Keep stirring until your mixture cools.

3 Pour the mixture into a clean, sterile, reusable jar.

This moisturising balm is gentle enough that you can use it on your entire body and your face too – a true all-purpose wonder-balm. It is non-greasy and sinks straight into your skin leaving it feeling perfectly nourished. If you are a minimalist with your skincare, you may find yourself needing only this balm for all of your skincare needs. It will even work well for removing make-up, softening your cuticles, dry elbows and knees.

Solid lotion bar

These solid lotion bars are so handy for travelling and are really nourishing and simple to use. Although I have tried to keep to the same ingredients for each recipe, this one introduces cocoa butter as it helps to keep the structure of the bar in solid form and is also very nourishing. Cocoa butter is a little heavy to be used on the face, so I would recommend using them only on the body.

> *Makes 2 medium bars*
> 53g mango butter or shea butter
> 45g cocoa butter
> 1g vitamin E oil

1 Gently melt together the mango butter and cocoa butter using the double boiler method until they are liquid and combined.
2 Set aside to cool to room temperature then add the vitamin E oil.
3 Add the mixture to a silicone mould and place in the fridge to chill overnight.

These bars should last up to 3 months. I love to take these with me when travelling and also find they are great for massaging my skin. They also make beautiful gifts and are always so well received by my family and friends.

Bath bombs

My children particularly love bath bombs, so I had to share this recipe. These are so simple and easy to make, and lots of fun to create with the children! I love that it works as both a fun activity and also makes bath time fun too. They are full of natural, skin softening ingredients and are gentle enough for sensitive skin. I like to keep these simple, adding botanicals like dried lavender, rose petals and calendula petals, but you can also add food colouring too, if you like. You'll need DIY bath bomb moulds; you can use silicone or aluminium moulds in any shape you choose.

> *Makes 4 large bath bombs*
> 200g bicarbonate of soda
> 100g citric acid
> 50g cornflour/cornstarch
> 50g Epsom salts
> 50g coconut oil or olive oil
> 10g dried botanicals
> Spray bottle containing water

1 Place the bicarbonate of soda, citric acid, cornflour and Epsom salts in a bowl and stir to combine.

2 Melt the coconut oil using the double boiler method until liquid.

3 Slowly add the oil to the dry mixture and stir to combine.

4 When fully combined, spray a small amount of water on the mixture and continue mixing together. You will only need to spray this once or twice to get the consistency you need.

5   Your mixture should appear like wet sand and start clumping together when it is ready for the mould.

6   Sprinkle the bottom of the moulds with the botanicals that you have chosen, then add the mixture to your moulds.

7   Add to your bath water and enjoy the lovely bubbles!

Eucalyptus

I add eucalyptus to my shower by tying a bunch behind the head of the shower and it has quite a few benefits for me. Similar to displaying flowers around my home, the eucalyptus visually brings me joy, but it also has some extra benefits that help to boost my mood and health.

When I take a shower, the steam helps to activate and release the essential oils of eucalyptus, providing an invigorating scent that can help reduce stress and promote relaxation and a positive mood. Eucalyptus is also known for its decongestant properties. Inhaling the vapours of the eucalyptus when the steam hits it can relieve symptoms of colds and sinus congestion.

I add a fresh bundle of eucalyptus to my shower every couple of months, I enjoy the aroma for the first few weeks when it is still fresh and when it has fully dried, I add a few drops of eucalyptus essential oil to the leaves to help it last longer and continue its aromatherapy benefits. Just a note when using essential oils: it is always beneficial to research each oil that you are using, particularly if you have children or pets in proximity to where you are using them. The eucalyptus is tied out of reach of my little ones.

I now grow eucalyptus in my garden so that I always have a fresh supply; it grows fast so you will find yourself pruning it often. Eucalyptus also looks fantastic with cut flowers as an extra bonus.

Natural face masks and skin treatments using upcycled ingredients

I like to use natural and upcycled ingredients to treat my skin. I love that it gives something a second use before it makes its way to the compost, and it also saves me money and reduces waste because I'm not buying extra products that just clutter up my bathroom. I have tried lots of recipes over the years and here are a few of my favourites.

Banana skins

The natural oils and antioxidants in banana peels can help to hydrate and soften the skin and can be used on a daily basis as a very mild yet nourishing exfoliant. The vitamins and minerals in banana skins can lighten dark spots and improve skin tone so are a powerful addition to your natural skincare routine if you suffer with hyperpigmentation.

To use your banana peels as a skin treatment, simply rub the inside skin of the banana peel all over your face after cleansing. Move in upward, circular motions and continue the process for about 1 minute, and leave the residue on your skin for a couple of minutes, then just wash away with warm water before moisturising.

Rice water

Rice water has been used for centuries for its beneficial properties for skin and hair. A rice water face mask contains vitamins and minerals that can help improve skin tone and reduce dark spots. The antioxidants in the rice water also help to soothe, hydrate and soften the skin. If you have acne-prone skin, you may also find that rice water helps to control excess oil production.

To use rice water on your skin, soak 1 cup of rice in 2–3 cups of water for an hour in the fridge before you use the rice for cooking. When straining the rice, make sure to keep the milky rice water that you would usually discard and store it in a clean reusable jar in the fridge.

To use the rice water as a face mask I like to dip my cellulose sponge into it and sweep it over my face in the morning and at night. I leave it to sit for a minute, usually while brushing my teeth, and then wash it off with warm water.

The rice water will keep for up to 1 week in the fridge and should be discarded after this time.

Yoghurt, honey and turmeric face mask

I always seem to find that I have a little bit of yoghurt left at the bottom of the jar, so whenever this happens, I whip myself up this face mask to nourish my skin – and it has lots of benefits!

The turmeric in the face mask helps to reduce inflammation and fight acne and is beneficial if you suffer from redness in your skin. The yoghurt is soothing and hydrating, and honey acts as a natural humectant, drawing moisture into the skin while also having antibacterial benefits.

To make this face mask, all you need to do is mix half a teaspoon of ground turmeric, half a tablespoon of yoghurt and half a tablespoon of raw honey together. Apply to your skin, relax for 10–15 minutes then wash away.

Sugar and upcycled coffee grounds body scrub

I love to save some of my coffee grounds to make this body scrub. This is a recipe that I make and use on the same day. It will keep for 2–3 days in the fridge but no longer, so make sure to use it all up!

Coffee grounds are great for the skin and as well as helping to remove dead skin cells to leave the skin more radiant, the grounds are also rich in antioxidants and contain anti-inflammatory properties, which help soothe the skin. Massaging the coffee grounds into your skin can also help to improve circulation and brighten the skin.

1 tablespoon sugar
1 tablespoon coffee grounds
1 tablespoon jojoba oil

1. Combine the sugar, coffee grounds and jojoba oil in a bowl.
2. Place in a clean, reusable jar and store in the fridge.

I use a generous handful of this body scrub whenever my skin is feeling dry or itchy and it always helps to return it to its glowing and nourished old self. It's also brilliant for helping to reduce the appearance of uneven skin tones and scars.

Sea salt hair spray

I like to use this sea salt hair spray for my hair to create beach waves and add texture when I'm leaving it to dry naturally in summertime. It gives that straight-off-the-beach look, while the oil and gel still nourishes my hair.

Makes 1 x 250ml spray bottle
15g sea salt
15g argan oil
200ml boiled water, cooled until warm

1 Add the sea salt and argan oil to the warm water.
2 Stir until all of the ingredients are combined and the salt is dissolved.
3 Transfer to a clean, sterile, reusable spray bottle.
4 Shake the bottle before each use to thoroughly combine the ingredients.

Linseed hair gel

This is an all-natural hair gel that is really easy to make and a great way to style your hair while also providing it with nourishment. It is brilliant for creating waves and curls naturally and the vitamin E in the flaxseed also nourishes your hair.

> *Makes 1 medium jar*
> 30g whole linseeds
> 200ml water

1. Place the linseeds in a small saucepan and pour over the water.
2. Bring the mixture to the boil over a medium heat, stirring occasionally.
3. Once the mixture reaches boiling point, reduce the heat and allow to simmer, stirring frequently so that the mixture doesn't stick to the bottom of the pan.
4. Simmer for about 10 minutes until the mixture thickens to a gel-like consistency.
5. Strain through a fine-mesh strainer or muslin cloth into a clean bowl to separate the seeds and the gel, then discard the seeds.
6. Allow the gel to cool and transfer to a clean, reusable container.
7. Use the gel on wet hair, gently scrunching it in.

You can store this in a container for about 5 days in the fridge. This DIY linseed hair gel is a natural, effective way to style your hair while avoiding the chemicals found in commercial hair gels. It is incredibly nourishing, and if you have any left over at the end of the week, you can use it as a hair mask.

MY PLASTIC-FREE ...

Botanical bath salts

These botanical bath salts are a beautiful way to relax and unwind. The fragrance is uplifting, and the ingredients help to soften your skin and soothe your body. The Epsom salts are soothing, and the kaolin clay has detoxifying properties which help to draw out impurities. They also make a fantastic gift to share with your friends and loved ones! I present them in pretty jars so they can be reused by their lucky recipient.

> *Makes 1 large jar or 2 medium jars*
> 600g Epsom salts
> 100g bicarbonate of soda
> 40g kaolin clay
> 10g botanicals of your choice (rose petals, lavender buds, chamomile or calendula petals)

1. In a large bowl, mix together the Epsom salts, bicarbonate of soda and kaolin clay.
2. Place a large scoop of the salt and clay mixture into a clean jar then sprinkle over a thin layer of the botanicals. Continue filling the jar in alternating layers until it is filled.

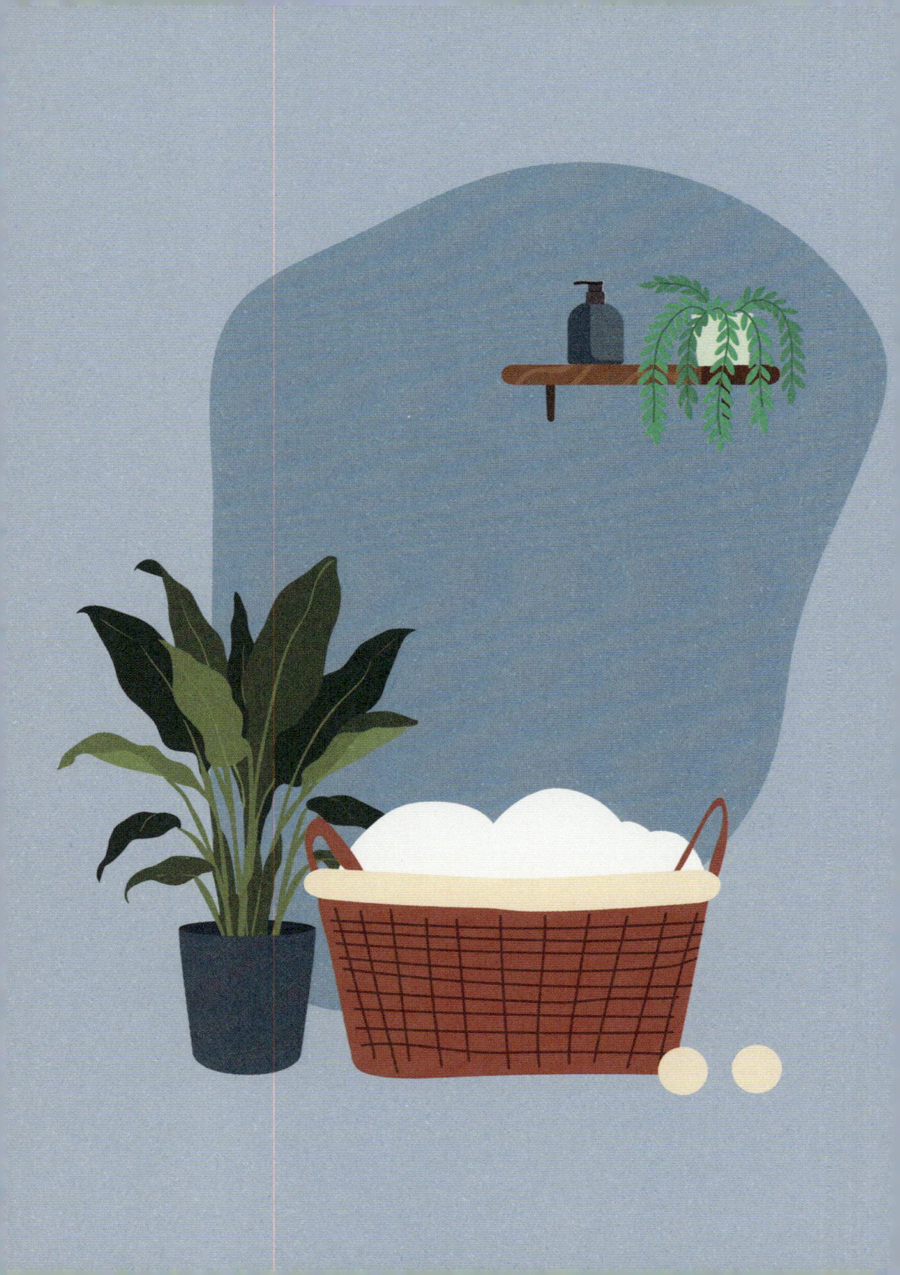

MY PLASTIC-FREE CLEANING

HEALTHY SUSTAINABLE SWAPS

Cleaning products – what a minefield they are! Over the past decade I have tried so many different cleaning methods, as cleaning products were one of the catalysts that led me to living a more waste- and plastic-free life. After a long and winding road to motherhood, I became acutely aware of what was going into my body. When I reflect on this, I find it really strange that being pregnant and looking after my daughter was probably the first time that I realised I also needed to look after my body too.

I began to question the products that I was using in my home and what they were doing to both me and my family. I distinctly remember while cleaning the shower one day, I developed such a headache that I needed to lie down. As I was pregnant at the time I began panicking that whatever I had inhaled may have affected my baby. This was the first time that I started to look into the effects of cleaning products, and I couldn't believe that I had been using products that were full of toxic chemicals that were harmful to myself and the environment without giving them a second thought. I certainly didn't want them in my home any more and needed to find alternatives as quickly as possible, so that I could clean my home safely and effectively.

Over the years I have tried and tested so many cleaning products and I now use a mixture of DIY and shop-bought, eco-friendly products in my home. The following are super-easy to create, so why not give it a go?

... CLEANING

Multi-purpose cleaner

For use on granite and quartz surfaces, stainless steel and glass

I wanted to make a cleaner that you could use all around the home and that is effective on every surface. We have granite surfaces in our kitchen that can easily look streaky, and no cleaning product seemed to make any difference until I started making this. The secret ingredient is rubbing alcohol, which is also known as isopropyl alcohol. I always make sure to buy isopropyl alcohol 70% as this means that it works effectively without evaporating too quickly. Isopropyl alcohol is something that is difficult to find plastic-free, so to make it low-waste I buy it in a 5-litre container and store it in a childproof cupboard in my kitchen. It lasts for years this way and means that I'm not purchasing lots of little plastic bottles.

I have included a reference at the back of the book for handy websites and products that I use to buy my raw ingredients (see Resources on page 252).

Makes 1 x 500ml bottle
150ml rubbing alcohol (70%)
300ml water
10 drops of lavender essential oil (optional)
10 drops of geranium essential oil (optional)
10 drops of basil essential oil (optional)
1 tablespoon liquid soap

1. In a jug, mix together the rubbing alcohol, water and essential oils, if using.
2. Lastly add the liquid soap and gently stir.
3. Transfer to a glass bottle and attach a trigger spray.
4. Label the bottle clearly.

Use this on all of your surfaces for a clean and sparkling finish! The additional drops of essential oil help it to smell amazing and also add extra benefits. Both the lavender and basil essential oils are antibacterial and antifungal, and they smell wonderful as you clean.

Glass cleaner

For use on glass and mirrors

This glass cleaner is going to rock your world! It's amazing at making your glass and mirrors completely streak free with no effort at all. The secret ingredient in this recipe is cornflour (also known as cornstarch). Cornflour helps to clean glass effectively due to its fine, abrasive texture which, when applied to glass, helps to gently scrub away dirt, grime and any residue without scratching the surface. It also has the ability to absorb moisture and grease, which helps to remove fingerprints, smudges and streaks.

Makes 1 x 500ml bottle
75ml rubbing alcohol (70%)
75ml vinegar

300ml water
10 drops of mandarin essential oil (optional)
10 drops of lavender essential oil (optional)
1 tablespoon cornflour/cornstarch

1 In a jug, mix together the rubbing alcohol, vinegar, water and essential oils, if using.
2 Lastly add the cornflour and gently stir.
3 Transfer to a glass bottle and attach a trigger spray.
4 Label the bottle clearly.

The additional drops of essential oil, if you have used them, help it to smell amazing and also add aromatherapy properties. Both the mandarin and lavender essential oils are also antifungal.

Toilet cleaner

One of the first things I wanted to get rid of in our house when making the change to eco-friendly cleaning products was the dreaded bleach. I can't believe that this was one of the main reasons cleaning the bathroom gave me a headache. Even when I made sure the room was well ventilated, it never occurred to me how harmful that must have been for my health! Additionally, bleach going straight into the toilet is so bad for the environment and is harmful for aquatic life, so it definitely had to go.

Knowing that I wanted to find an alternative for bleach was the first step, but finding the alternative took me some time. Years ago, I started working with an eco-friendly company that produced the raw ingredients for cleaning and that is where I discovered citric acid. Citric acid is a versatile and effective natural cleaner that is excellent for dissolving the limescale build-up that you often find in your toilet.

> 250g citric acid

1 Using a toilet brush, plunge your toilet a few times to lower the water level.
2 Pour hot water (not boiling) into the toilet.
3 Add the citric acid to the toilet bowl and leave to sit for a few hours.
4 The citric acid will now have eaten the limescale and naturally disinfected your toilet. I then clean the toilet bowl and under the rim with my multi-purpose cleaner (see page 116).

Cream cleaner

Multipurpose cleaning scrub

I always have a jar of this cream cleaner in my cupboard. It's so incredibly simple to make and can be used all around your home in many different ways. I use it to remove marks on the surfaces in my kitchen, to remove soap scum in the bath and shower, in fact, to remove any grimy surfaces that need a little elbow grease. The cream cleaner is a simple mixture of liquid soap (washing up liquid) and bicarbonate of soda.

Although we generally only think to use it when washing dishes, liquid soap is a gentle but effective cleaner that can be used all around the home. It is also well known to reduce the spread of viruses. (It is no surprise that during the coronavirus pandemic healthcare advice was to use liquid soap to kill germs in preference to using hand sanitiser.) In addition to liquid soap, bicarbonate of soda is another versatile and effective cleaning product. It is a mildly abrasive cleaner and effective for scrubbing surfaces without scratching them. It is also a really handy product for cutting through grease on stovetops, ovens and cooking utensils.

Makes 1 x 300g jar
150ml liquid soap
140g bicarbonate of soda

1 Mix together the liquid soap and bicarbonate of soda in a bowl.
2 Add to a clean, reusable jar and label the jar clearly.

You will find that this little wonder scrub is really tough on stains around your home – marks on your kitchen surfaces, like tea or turmeric stains, are lifted in an instant! It's also fantastic to use in the bathroom and helps to tackle limescale and soap scum. I also use it to clean my oven naturally!

. . . CLEANING

Wood polish

For cleaning kitchen cabinets, skirting boards and wooden furniture

I have found through personal experience that when using natural cleaning products, you need to be careful when cleaning wood. Wood is a very porous surface, and often has a finish that can react badly to both very acidic and very alkaline products. I make a batch of this wood polish, which I use on my skirting boards, painted wood surfaces, sideboards and tables, and it helps to both clean the surface and repel dust particles, meaning that you won't need to clean the surfaces as frequently – hurrah!

> *Makes 1 x 200ml spray bottle*
> 60ml walnut oil
> 30ml white vinegar
> 100ml water
> 10 drops of lavender essential oil
> 10 drops of lemon essential oil

1 Mix all the ingredients together in a jug.
2 Pour into a clean, reusable spray bottle and label the bottle clearly.
3 Spray on wooden surfaces to kill dust mites and repel dust.

. . . CLEANING

NATURAL CLEANING POWDERS

I use natural cleaning powders in conjunction with homemade cleaning products for many tasks all around the house. Here, I list each cleaning powder and what you can use them for, as well as including the limitations of the powders so that you don't make any cleaning errors along the way.

Natural bleach

Natural bleach, also known as sodium percarbonate, oxygen bleach or green bleach, is an effective cleaning product that can be used for both laundry and household cleaning. It is one of my favourite 'wow' products because it is so effective. When dissolved in hot water, sodium percarbonate releases hydrogen peroxide and soda ash, providing effective cleaning, bleaching and disinfecting properties. I find it particularly useful for laundry, but you can also use it as an all-purpose cleaner while it is active (see opposite).

Laundry

To use natural bleach as a clothes brightener and deodoriser, add 30g to your normal washing. To use natural bleach as a pre-soak add 60g to 5 litres of hot water and leave your washing to soak for a few hours. Then add to your normal washing.

Cleaning hard surfaces

Natural bleach can work effectively to clean tiles, grout, countertops, sinks, bathtubs, to mop floors, clean bins and disinfect bathroom surfaces. It can also be used to clean outdoor decking, furniture and patios.

To make a natural bleach multi-surface cleaner add 2 tablespoons of sodium percarbonate to 500ml of hot water and stir to dissolve. Use gloves and a sponge or scrubbing brush and apply to the surfaces that you are cleaning. The solution will be active and foaming, which is the best time to use the mixture as this is what makes it so effective. DO NOT add this active mixture to a spray bottle. I find a Pyrex jug is sufficient.

When not to use natural bleach
Natural ingredients are extremely effective for cleaning, but they do have their own rules. It's important to note that you should not use natural bleach on delicate fabrics, such as wool, silk or leather. Do not use them on aluminium surfaces, as sodium percarbonate can react with aluminium, causing it to discolour or corrode.

Citric acid
The acidic nature of citric acid makes it effective for a wide range of cleaning tasks, especially those that involve limescale, soap scum and rust.

To make a citric acid solution for cleaning add 50g to 500ml of hot water. You can then either apply this solution with a sponge, add it to a spray bottle or soak your affected objects in it by submerging them.

Limescale removal
Citric acid is great for removing limescale build-up in kettles, coffee makers, dishwashers and faucets. It can also be used to clean shower heads where hard water deposits accumulate.

Rust removal
Citric acid literally eats through rust and can remove it quickly and efficiently leaving no trace behind.

Bathroom cleaning
Citric acid is great for cleaning soap scum, mineral deposits and as a general surface cleaner too.

When not to use citric acid
Citric acid should not be used on marble, granite or natural stone surfaces, as the acid can corrode these and cause permanent damage. Also avoid using it regularly on wood with polished or wax finishes as it can strip the coating. Always test on a small, inconspicuous area before using on a larger surface to ensure compatibility.

Soda crystals
Soda crystals have been used for centuries as a natural cleaning product. They are closely related to bicarbonate of soda but have a much stronger cleaning power due to their high alkalinity, making them effective at breaking down grease, oil or tough stains.

To make a soda crystal solution for cleaning add 50g to 500ml of hot water. You can then either apply this solution with a sponge or add it to a spray bottle. Use this for:

- Degreasing kitchen surfaces, cookware, extractor hoods, cupboards and ovens.

- Unclogging drains.
- Removing soap scum, water deposits, mildew and grime from the bath, shower, toilet, sink and tiles.
- Cleaning appliances such as washing machines and dishwashers.
- Breaking down dirt, moss and algae in the garden. It's useful for cleaning patios, garden furniture and paths.

When not to use soda crystals

Soda crystals are highly alkaline and can corrode or discolour aluminium after prolonged use. Also, avoid using on:

- Natural stone surfaces like granite, marble and slate, as they can strip the sealant or cause etching. I prefer to use my multi-purpose cleaner for this job.
- Lacquered, painted, waxed or polished surfaces as they can strip the finish and leave them looking dull or damaged.

Always test on a small, inconspicuous area before using on a larger surface to ensure compatibility.

Bicarbonate of soda

Bicarbonate of soda is mildly abrasive but a very effective cleaning powder that can be used for a number of tasks around the home, including:

- Deodorising – Bicarbonate of soda absorbs and neutralises odours, making it ideal for use in the fridge, for shoes, carpets and drains. Sprinkle it where you need it!

- Cleaning surfaces – It acts as a mild abrasive to clean countertops, sinks and stovetops without scratching them. I like to make it into a paste with liquid soap (see page 119).

- Polishing silverware and chrome – A paste of bicarbonate of soda and water can be used to polish any silverware, removing tarnish and restoring shine. It is also effective for cleaning chrome faucets, shower heads and sinks and can help remove water spots and mineral deposits.

- Carpet cleaning – You can sprinkle bicarbonate of soda onto carpets before vacuuming to absorb odours and lift dirt from fibres.

When not to use bicarbonate of soda

Bicarbonate of soda is considered safe for most surfaces, but prolonged use can cause wear due to its mildly abrasive nature. Always test on a small, inconspicuous area before using on a larger surface to ensure compatibility.

. . . CLEANING

MY PLASTIC-FREE . . .

BUYING ECO-FRIENDLY PRODUCTS vs MAKING YOUR OWN

Antibacterial multi-surface cleaner

In addition to making my own products, I have bought and used a good, eco-friendly, antibacterial multi-surface cleaner. This was a habit that I picked up during Covid when germs were well and truly on my mind, and I sent myself a little crazy cleaning. Thankfully, I was able to source an antibacterial multi-purpose spray refill that was eco-friendly and plastic-free and I used it for those high-touch areas where I wanted that little bit more peace of mind. See Resources on page 252 for brand recommendations.

I include this anecdote not because I think that DIY cleaning products are insufficient (they clean my house beautifully). There are a number of ways to create eco-friendly habits and it's up to you which you choose. It is not for me to tell you that using and making your own eco-friendly cleaning products from powders and raw ingredients is more eco-friendly than buying ready-made products from an eco-friendly business that uses non-toxic ingredients. You could even mix the two, which I sometimes do.

Again, I share a list of eco-friendly cleaning brands in the Resources on page 252 if this is something that you would prefer to do over making your own products.

Laundry detergent

I've used a few different methods for cleaning my laundry over the years. I have found a simple DIY recipe that works very well, and I also recommend a number of eco-friendly products from small businesses if you would prefer to buy your laundry detergent.

There are so many options out there if you want to go eco-friendly with your laundry, from a refillable 'egg' that washes your clothes without the need for detergent, to laundry sheets that you can add to your washing machine, and even making your own detergent. Here is a little more information about each option:

Ecoegg

An Ecoegg is a small, egg-shaped, reusable container filled with two types of mineral pellets that work together to clean clothes. It doesn't contain any harsh chemicals or fragrances and is suitable for sensitive skin. The egg itself is reusable and really convenient. All you need to do is pop it in the washing machine with your laundry. There is no need to measure and pour detergents, which really simplifies the process.

Laundry sheets

These are now a very common alternative to those laundry pods that contain lots of chemicals. They are basically sheets of detergent that you add to your washing machine for each wash. When combined with water and agitated in the washing machine, they dissolve into a detergent and clean your clothes. They are available in different eco-friendly fragrances, and you can also find them fragrance-free too. All the goodness and no nasties.

Laundry liquid

Refillable laundry liquid is also an option if you live close to an refill shop. Just take a container and fill it with as much as you need! There will be different fragrances for you to choose from, and it is all made from eco-friendly ingredients that won't harm the planet.

Homemade laundry detergent

You can also make your own laundry detergent, and it is really simple to do! Here is a recipe that I have found works great for your clothes and is really cost-effective too.

> ***Makes 1.2kg***
> 100g fine sea salt
> 100g Epsom salts
> 500g soda crystals
> 500g bicarbonate of soda

1. Mix all of the ingredients together in a bowl and store in an airtight jar.

2. Add 2–3 tablespoons of the detergent to each wash depending on your load (for heavily soiled items you may need to add a little more). You will find that the salts and bicarbonate of soda help to brighten whites and remove odours, as well as keeping colours vibrant, while the soda crystals help to remove tough stains and soften the water in your washing machine. As discussed with sodium percarbonate, you can also add a tablespoon of this powder for laundry that needs a boost and extra stain removal.

Fabric softener

There are two methods that I use for fabric softener, and both work well, likely because they both have the same properties. If you have never tried vinegar or citric acid as a fabric softener, then it may take a minute to get your head around this idea! But I promise you it works a charm.

Vinegar

It might surprise you that vinegar is a well-known and much-loved natural substitute for fabric softener. It helps to soften your clothes and reduce static, and adds a freshness to your washing. Simply add 60ml of white vinegar to the fabric softener dispenser of your washing machine or during the rinse cycle.

Vinegar also helps to break down detergent and mineral build-up in your clothing that can often make it feel stiff. It is especially useful in hard water areas where mineral build-up can really affect the texture of fabric. There is no need to worry about the smell of vinegar on your clothes either – it completely dissipates during the rinse cycle and just leaves your clothes smelling fresh and clean.

Citric acid

Using citric acid as a fabric softener is another natural alternative that offers similar benefits to vinegar. Here's how you can use citric acid effectively. Dissolve a tablespoon of citric acid into 60ml of hot water and allow it to cool. This is an important step when using citric acid on cold cycles, but the citric acid could be added straight to the drawer of your washing machine if you are washing your clothes at a warmer temperature. Add the citric acid solution to the fabric softener dispenser of your washing machine or during the rinse cycle.

Citric acid works very similarly to vinegar, but in my experience is more cost-effective. Like vinegar, it helps to soften fabrics by breaking down mineral deposits and detergent residue that can leave your clothes feeling quite stiff.

Air-drying your laundry and natural stain removal
Air-drying, or line drying laundry is such a simple and effective way to reduce energy consumption over using a tumble dryer. It is also gentler on the fabrics that you wash and reduces wear and tear, helping them last longer. It is a really simple way to be more eco-friendly and extend the life of your clothing.

Did you know that sunlight is also a really effective stain remover? The sun's ultraviolet rays break down the molecular bonds in certain stains like food, sweat and grass. This is particularly effective with white clothes. To enhance the effect on white clothing you can pre-treat the stain with a little vinegar or citric acid solution.

Dryer balls
If you are in a pickle and have not the time or capacity to air-dry your clothes, dryer balls are a great addition to your laundry to make the process more eco-friendly. When added to your drying, dryer balls separate clothes and allow more air circulation in the dryer, reducing drying time significantly, making the process more energy-efficient and more cost-effective too.

Stain remover bars
I also have a handy stain remover bar in my laundry products. Stain remover bars are solid soap bars designed to help remove stains from fabric. They are an eco-friendly alternative to stain remover sprays and products.

To use the stain remover bar, I just wet the bar and place it on the area of fabric where there is a stain and rub the bar gently onto that area in circular motions. You'll notice that a foam will form and then you can use your fingers or a soft brush (like a toothbrush) to gently scrub the stain. For tougher stains, after doing this, I let the soap sit on the stain for 5–10 minutes and then add that item of clothing to the wash.

These bars are portable, and I often take them on holiday with me to do spot washes and small hand washes when needed.

Cleaning the oven

You are never going to need to use toxic oven cleaner again after you have tried this method. It works an absolute charm!

> 500ml water
> 250ml vinegar
> 100ml DIY cream cleaner (see page 119)

1. Pour the water and vinegar into a Pyrex/oven-safe bowl and place on the bottom shelf of the oven.

2. Turn the oven on high and allow it to heat up for 15 minutes. The steam created helps to loosen any food residue, and the vinegar will help to clean and degrease the oven.

3. Switch off the oven and allow to cool for 5 minutes without opening the door.

4. After 5 minutes, open the door and, wearing gloves, use a sponge to wipe the cream cleaner all over the oven including the door, racks and sides. Try to do this as swiftly as possible so the oven is still hot. Close the door again and leave for another 15 minutes until the oven has cooled.

5 You can now open the oven door again and, using an abrasive sponge, clean away the residue and leave your oven spotless.

If your oven is overdue a clean, I find that a glass scraper works wonders on the oven door at this point, and the racks can also be left to soak in a bath or sink overnight covered with hot water and a few tablespoons of soda crystals. if you are using this method, I would recommend placing an old towel underneath the oven racks so that there is no discolouration on the bathtub.

Cleaning the microwave

I clean the microwave using a similar method to cleaning the oven. This is such a simple method. I guarantee you will be so shocked by how effective it is and hardly any effort is needed!

500ml water
250ml vinegar

1 Pour the water and vinegar into a microwaveable container and place in the microwave.
2 Microwave for 5 minutes on high so that the mixture heats up and steams the microwave. The steam helps to loosen anyfood residue, and the vinegar will help to clean and sanitise the microwave.
3 Allow to cool for 5 minutes without opening the microwave door.
4 After 5 minutes, open the door and wipe the inside of the microwave.

... CLEANING

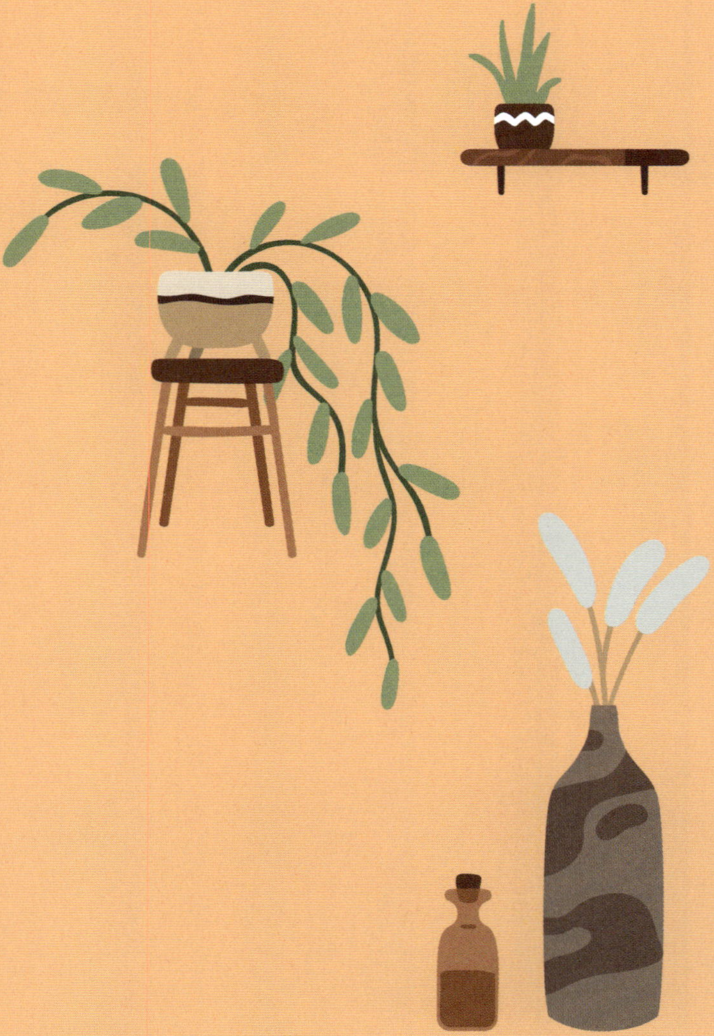

MY PLASTIC-FREE LITTLE LUXURIES

THE JOY OF WASTE-FREE LUXURIES

I wanted to share this chapter in the book to show all of the ways that we can reduce waste but enjoy those little luxuries at the same time. Being eco-friendly is so often viewed as boring and bland – I couldn't agree less, and I want to be able to show you that it can be quite the opposite!

I find the process of making and creating so rewarding and calming. It's something that helps the planet, but also helps your soul too. So, make sure you take the time to create a little luxury.

Candles

I like to use wax melts and candles in my home to create a warm and cosy atmosphere in the evening. There is something about a warm flame that I can simply sit and watch. There are so many options available in the shops and there are a few important things to consider when buying wax melts and candles for your home.

If the candles seem cheap, they are very likely made from paraffin wax, which is a petroleum by-product, also known as crude oil – a fossil fuel. It may surprise you that the majority of candles are made from paraffin. Burning paraffin has been found to release VOCs (volatile organic compounds) and also black soot when it is burned, which is not good for our health.

I make sure to look for natural wax (coconut wax, soy wax or beeswax), candles and wax melts as these are the most natural available. It's also important to look at the fragrance that is present in the candles as a lot can be synthetic. Looking for candles that are made with essential oils and natural fragrance is really useful to avoid the synthetic alternatives.

You can also make your own candles, and they are a lot simpler

. . . LITTLE LUXURIES

143

than you may think. I am lucky enough to know a local beekeeper who has gifted us leftover beeswax from the hive, which I have been able to filter and make candles with. I also upcycle candles from the odds and ends that I save as they become unusable.

DIY upcycled candles

Upcycling old candles to make new ones is a great way to reduce waste and create something useful and beautiful from something that you would normally throw away. And it's so simple and satisfying to do. I have a large jar that I like to add old and broken candles to. When the jar is full, I melt them down and turn them into a new one.

> Old and broken candles
> Pre-waxed candle wicks
> Upcycled jars for new candles to be poured into

1 Gather all your old and broken candles until you have enough to melt into a new one.

2 Remove any debris from the candles then melt the wax pieces together using the double boiler method.

3 While the wax is melting, prepare the jars that you are using to make your candles by attaching the new wick to the bottom of the jar using a blob of melted wax or compostable tape.

4 Use pencils to keep the wicks centred in the jars.

5 When the wax is fully melted, remove it from the heat and carefully pour into the jars – pour slowly to avoid bubbles.

6 Set the jars aside to allow the wax to cool and set.
7 When the wax has fully set, trim the wick to about 5mm above the wax surface.

DIY dipped meditation candles

I really enjoy lighting a meditation candle in the morning and also when practising yoga. I find that they help me to focus, and bring mindfulness and relaxation. The gentle, flickering light also promotes tranquillity.

Making your own dipped meditation candles is also a very relaxing project, one that you could do alone or even with friends to enjoy over a craft evening.

> Soy wax or beeswax
> Candle wicks

1 Melt your chosen wax using the double boiler method, until it is completely liquid and then remove from the heat. I use a stainless-steel pot that is tall and slim as the top part of the double boiler, as it allows for the candles to be dipped.
2 Cut the wicks to your desired length, leaving extra length for holding and dipping the wicks.
3 To dip the wicks, hold the wick straight and dip it into the melted wax, then lift it out.
4 Allow the wax to cool and harden for a few seconds before dipping it again.

5 Repeat this process of dipping and cooling the candles, until the candles reach the desired thickness. This could take 20–30 dips and is a therapeutic process.

6 After the final dip of your candles, hang them to cool and harden completely. Make sure that you hang them straight to avoid the candles bending.

These are small in size, and a great project for beginning to make your own candles. When you gain confidence making these candles you can make larger candles of any size you want.

Room fresheners

I love my home to smell beautiful. Even if it's a little untidy, a gorgeous smelling home amidst all the chaos just makes sense to me and helps me feel relaxed. The smell of our homes can have a significant impact on our moods and overall wellbeing. Different fragrances can evoke memories and even make our homes feel more comfortable and inspiring to us. Here are a few different methods I use to fragrance my home naturally.

Diffuser

The first method is probably the simplest and something that I like to do every day as routinely as I make a morning coffee. I add essential oil to a diffuser to gently fragrance my home. The diffuser that I have has a ceramic top, which glows a warm yellow and shoots out a little steam into the air all day long, which I find really calming. These are my two favourite diffuser blends:

Lavender and geranium
6 drops of lavender essential oil
6 drops of geranium essential oil

Mandarin and rosemary
6 drops of mandarin essential oil
6 drops of rosemary essential oil

MY PLASTIC-FREE . . .

Simmer pot

A DIY simmer pot is an easy and natural way to create a cosy and inviting atmosphere in your home. It's something I love to do in the evening and on special occasions. Simmer pots are especially popular in our home at Christmas time, but I will also share a recipe for a summer simmer pot using seasonal ingredients that you can probably find in your garden or pick up from your local grocery shop.

Summer simmer pot

> 1 lemon or orange
> 2 sprigs of fresh rosemary
> Fresh mint
> 1 star anise
> 1 cinnamon stick
> A small handful of lavender (fresh or dried), rose petals or jasmine

1. Slice the citrus fruit and crush the herbs to release the aromas.
2. Half-fill a medium pan with water – you want enough to prevent the pan from drying out too quickly as it simmers.
3. Add the ingredients to the pan and place it on the stove.
4. Bring to a gentle simmer over a low heat – do not let it boil vigorously as this can cause the water to evaporate too quickly.
5. Check the pan regularly and top up with more water if necessary, to keep the ingredients covered.

Winter simmer pot

> 1 orange
> A couple of sprigs of fresh or dried rosemary or a couple of sprigs of pine or bay leaves
> 1 cinnamon stick
> 1 teaspoon cloves
> 1 star anise
> ½ teaspoon nutmeg
> A small handful of cranberries or fresh ginger

1. Slice the orange and crush the herbs to release the aromas
2. Half-fill a medium pan with water – you want enough to prevent the pan from drying out too quickly as it simmers.
3. Add the ingredients to the pan and place it on the stove.
4. Bring to a gentle simmer over a low heat – do not let it boil vigorously as this can cause the water to evaporate too quickly.
5. Check the pan regularly and top up with more water if necessary, to keep the ingredients covered.

Once you have used your simmer pot, you can keep the ingredients in the fridge for a couple of days and reuse them. Another great way to make a simmer pot is to use a slow cooker, which is a less energy-intensive method, just leave the lid off and let the aroma fill the air.

Room spray

This is a really simple recipe. I like to add essential oils and tend to stick to the same blends that I use in my diffuser. The diffuser creates a lovely fragrance that lasts throughout the day, whereas a room spray adds a quick shot of beautiful fragrance when you feel like your room needs a freshen up. As discussed in the DIY cleaning products section, rubbing alcohol is something that I buy in a large container and keep in a child-proof cupboard under my sink because I use it quite often with both room spray and my cleaning products.

> 150ml water
> 45ml rubbing alcohol (70%)
> 35 drops of an essential oil blend (see below for a list of my favourites)

1. Mix the ingredients together in a bowl.
2. Store in a spray bottle and use within 3 months.

Essential oil blends

- Spring – During the spring I like to use rejuvenating citrusy and herbaceous fragrances as it just brings a freshness to my home. My favourite blend for springtime is lemon, mandarin and rosemary.

- Summer – I like to use bright, floral, citrus and fresh blends during the summer months. My favourite blend is lavender, geranium and lemon.

- Autumn – For me, autumn essential oils are rich and comforting with a little bit of spice. I love to naturally fragrance my home with these cosy smells, which make me feel like hibernating! My favourite blend is cinnamon bark, ginger, sweet orange and clove.
- Winter – I love to use cosy, warm, spicy and woody essential oils throughout winter. My favourite blend is cinnamon bark, clove, pine needles and mandarin.

Dream spray is a little additional room spray that I use in my daughter's bedroom. She finds the fragrance really settling and it helps her to drift off to sleep. It's also really nice to spray on your pillow at night. It's a blend of lavender, geranium and clary sage.

. . . LITTLE LUXURIES

Bathroom room spray

This is another very simple recipe that I make and keep in the bathroom for any moment that I find it needs a little freshen up! Lemongrass is excellent for combatting odours, which is why I always use it in this bathroom blend. I tend to make a 200ml spray bottle and make a fresh batch every month.

> 150ml water
> 45ml rubbing alcohol (70%)
> 15 drops of lemongrass essential oil
> 10 drops of lavender essential oil
> 10 drops of geranium essential oil

1. Add all of the ingredients to a jug and stir to combine.
2. Pour into a clean, sterile, reusable bottle with a spray top.
3. Clearly label the bottle.

You can play around with any essential oils you like for this recipe, but I recommend keeping the lemongrass fragrance as it is super-effective in the bathroom and blends so well.

REMEDIES FROM NATURE

When I began my journey to reduce the waste and plastic in my life, it opened up a whole new world to me, and how we are connected and rooted in nature has become such a huge part of that.

When you realise the impact that you are having on the world, and indeed the impact of the products that you are using, it makes you want to explore natural alternatives that you can use in your home, which in turn opens up a world of possibilities.

Here are some of my favourite herbal remedies, which I make to keep our family healthy and also strengthen that connection with nature that we so often feel is lost.

Fire cider

This isn't for the faint-hearted! This fire cider is a potent herbal remedy made by infusing various herbs, spices and vegetables. I like to take a tablespoon of fire cider every day during the cold season to improve my immunity and fight colds and flu. In fact, whenever I feel like I'm coming down with something, a quick shot of fire cider tends to sort it out!

Makes a 1-litre jar
100g fresh horseradish root, grated
100g onion, chopped
50g garlic cloves
50g fresh ginger root, grated
50g lemon, sliced
1 chilli pepper, chopped
1 tablespoon dried rosemary leaves
1 teaspoon/thumb fresh turmeric
A pinch of black pepper
A pinch of ground cinnamon
Zest and juice of 1 lemon
Apple cider vinegar (to fill to the top)
1 tablespoon raw honey, to taste

1. Put the horseradish root, onion, garlic, ginger root, sliced lemon and chilli into a large 1-litre capacity jar.

2. Add the herbs, spices and lemon zest and juice.

3. Pour over the vinegar until the ingredients are completely submerged, leaving 3cm of space at the top of the jar.

4. Cover the jar with waxed paper or a muslin cloth to stop the vinegar corroding the lid of the jar, then close the lid.

5. Place the jar in a cool, dark place for up to 6 weeks to allow the contents to infuse, shaking the jar daily to help the process.

6. After 6 weeks, strain through a fine-mesh strainer or a sieve lined with a muslin cloth. Press down the mixture to extract as much liquid as possible.

7. Add honey to taste.

8. Transfer to a clean, sterile reusable jar.

Apple cider vinegar

Apple cider vinegar is used as a remedy for bloating or indigestion, and a natural remedy for sore throats. The acetic acid in apple cider vinegar helps to activate digestive enzymes and, while it is acidic, it actually has an alkalising effect on the body when metabolised, which can help with conditions caused by excess acidity.

Now I have shared some of the benefits of apple cider vinegar, I thought I would share with you how you can make your very own from apple scraps.

> 1 tablespoon sugar
> 300ml water
> Enough apple cores and peels to fill a 500ml jar halfway
> (these can be frozen until you have enough)

1. In a jug, dissolve the sugar in the water.
2. Place the apple scraps into a jar until they have filled it halfway.
3. Pour the sugar water over the scraps until they are fully submerged.
4. Cover the jar with a muslin cloth and secure with a rubber band. Store in a cool, dark place for up to 3 weeks. Stir the mixture each day to prevent mould from forming.
5. After 3 weeks the mixture should smell like vinegar. At this point you can strain out the apple scraps using a fine-mesh strainer or sieve lined with a muslin cloth and compost them. Pour the liquid back into the jar and cover again with a muslin cloth.
6. Allow the vinegar to ferment for a further 3–4 weeks, stirring occasionally. The longer it ferments, the stronger the apple cider vinegar will be.

Store your vinegar in a clean, sterile, reusable jar. You can keep this indefinitely – over time, you may notice a gelatinous substance form in the vinegar – this is known as the 'mother' and should be celebrated as it means the fermentation process is doing its thing! The mother contains beneficial bacteria that can aid in digestion and further promote gut health. The mother can also be used to make a new batch of apple cider vinegar.

Immunity bombs

I always have some immunity bombs in my freezer and whenever I'm feeling tired or need a little boost, I pop one into a cup of hot water and drink it like a warm tea. These bombs are also a fantastic way of naturally boosting your immunity, aiding your digestion and filling your body with antioxidants – and they taste great too!

> 4 lemons
> 2 apples
> 1 hand of ginger
> 2 tablespoons ground turmeric
> A pinch of cinnamon
> A pinch of pepper

1. Peel the lemons and core the apples.
2. Add the lemons, apples, ginger, turmeric, cinnamon and pepper to a blender and blitz to a thick liquid consistency.
3. Pass the liquid through a fine-mesh strainer to remove the pulp then pour the liquid into an ice-cube tray.
4. When you would like to make your tea, take a cube, add to your favourite cup and pour over hot water.

If you would like to keep the pulp then you can skip the straining steps in the method, the choice is yours. I personally like to separate the pulp and enjoy adding it to a powerful green smoothie that gives me an extra boost when I need it.

. . . LITTLE LUXURIES

Wellbeing gummies

I share a number of recipes to help boost immunity and wellbeing in this book, and as important as it is for me, it's also important for my children to have that boost too. I love to make wellbeing gummies for them, and these elderberry ones are one of my favourites.

> 80ml cold water
> 3 tablespoons unflavoured gelatine powder, or agar agar as a vegan alternative
> 250ml elderberry syrup (see page 189)
> 1 teaspoon lemon juice

1. Pour the water into a small bowl add the gelatine and let it sit for a few minutes to bloom.
2. Gently heat the elderberry syrup and lemon juice and, once warm, add the bloomed gelatine to the pan.
3. Stir thoroughly until the gelatine is dissolved – this should take about 2 minutes.
4. Pour into some gummy moulds. It may be helpful to use a dropper to do this.
5. Place the moulds in the fridge and leave to set for a few hours.
6. When set, carefully pop the gummies out of the moulds. If you find they are a little difficult to remove, freeze them for a few minutes to make them more solid. (But not for too long as this will alter the texture of the gummies.)

7 Store in an airtight container in the fridge where they should last for up to 2 weeks.

These homemade elderberry gummies are not only delicious but also provide the immune-boosting benefits of elderberries. If you have children who prefer a sweeter gummy, you can also add a little honey to the mixture before pouring into the moulds.

Herbal teas

Did you know that many of the tea bags that we use contain microplastics? I didn't either until a few years ago. There are a few brands of tea bags that I know and trust that are plastic-free (see Resources on page 252), but I also love to make herbal teas and infusions from scratch to extract all of the fresh, organic compounds from the herbs.

The process of making and drinking tea can be a calming ritual, providing a moment of mindfulness and relaxation in a busy day. I try to make a pot of tea every morning, sometimes if it's hot I turn this into a cold infusion. I'll share both methods with you, including the herbs that I like to grow to make my teas.

> 'There is something in the nature of tea that leads us into a world of quiet contemplation of life' – Lin Yutang

Botanicals to grow in your garden for teas:

Rose is such a soothing tea that has a delicate floral flavour and can help with stress relief and digestion.

Lavender is a calming herb known for its soothing properties. It can ease headaches, support digestion and lift your mood.

Peppermint is refreshing and invigorating, known for its ability to relieve bloating and indigestion. It also helps to ease respiratory issues and headaches, and the refreshing aroma is said to improve mental clarity.

Lemon balm is a soothing herb known for its calming properties. It helps to promote relaxation and reduce stress, and supports your digestion.

Chamomile is one of the most popular herbs, known for its delicate floral favour and calming effects. Chamomile helps to reduce stress and anxiety, and promotes better sleep. It's also said to reduce inflammation, support your overall health and relieve menstrual cramps.

Rosemary is a really aromatic herb known for its pine-like flavour. It is known to improve cognitive function and memory, and can help relieve indigestion. The aromatic compound is also said to reduce stress and anxiety.

... LITTLE LUXURIES

Citrus fruits that you can add to your botanicals:

Lemon – drinking hot lemon water is a great way to start the day. It helps to keep you hydrated, it's rich in vitamin C, which supports the immune system, and can even flush out toxins and support your kidney function.

Lime – lime is tart with a slightly sweeter edge than lemon and is also high in vitamin C and promotes hydration. Limes are packed with antioxidants that protect cells from inflammation. I like to add lime zest to my teas to give them a lively twist.

Orange – is also rich in vitamin C and antioxidants. It can add a natural sweetness to herbal tea and pairs well with chamomile, mint, ginger and cinnamon.

Spices that pair well with your herbal teas:

Cinnamon is warm, sweet and slightly spicy. It has anti-inflammatory properties and can help to regulate blood sugar levels.

Ginger is spicy, dusty, warming and aids digestion. It reduces nausea and has anti-inflammatory effects too. It complements herbal teas like lemon, mint or turmeric.

Cardamom is citrusy and slightly spicy. It is good for digestion, relieves bloating and has antimicrobial properties.

Cloves have antioxidant and antibacterial properties and can help with respiratory and digestive issues.

Turmeric is known for its anti-inflammatory and antioxidant effects. It also boosts immune function.

Nutmeg is warm, nutty and sweet and can help with digestion and stress relief.

Star anise is sweet and has a liquorice taste. It's known for its antibacterial and antifungal properties, and can support digestion.

Peppercorns are spicy, sharp and have anti-inflammatory properties, and enhance digestion too.

Fennel seeds are sweet and slightly liquorice-like. They help with digestion, reduce bloating and soothe the stomach.

To make a warm herbal tea from fresh ingredients:

1. Wash your chosen botanicals and lightly crush the leaves with your hands to release their natural oils. Add to a small pan or a mug.
2. Pour over boiling water and steep for 5–10 minutes, depending on how strong you would like the flavour.
3. Add any citrus fruits and spices that you desire.
4. Strain out the loose leaves.
5. Add a little honey or maple syrup, to taste, if you would like a sweeter tea.

To make a cold infusion herbal tea:

1 Wash the botanicals that you are going to add to your tea and lightly crush the leaves with your hands to release their natural oils.

2 Add your botanicals and any citrus fruits and spices to a pitcher or jar and top up with cold water. Allow to infuse for 4–8 hours, depending on how strong you want the flavour.

3 After the infusion period, strain out the loose leaves.

4 Add a little honey or maple syrup, to taste, if you would like a sweeter tea.

To make sun tea:

Another way to make cold tea is by a method called sun tea. It is a way of brewing tea using sunlight, and is a simple, energy-efficient way to make iced tea. It can enhance the flavours of the tea due to its slow infusion process.

1 Wash the botanicals that you are going to add to your tea and lightly crush the leaves with your hands to release the natural oils.

2 Add your botanicals to a large jar and top up with cold water.

3 Cover the jar with a lid or clean cloth and place in a sunny spot, such as on a windowsill, where it can get direct sunlight.

4 Steep the tea in the sunlight for 2–4 hours, depending on the strength that you would prefer.

5 After the infusion period, strain out the loose leaves.

6 Add a little honey or maple syrup, to taste, if you would like a sweeter tea.

FORAGING GIFTS

There is so much joy to be found in the pleasures of gathering and foraging your own food. It is something that is so often overlooked in today's times, where common fruit and vegetables appear on supermarket shelves all year round despite the air miles and manufacturing process it takes for them to arrive there – not to mention all that plastic packaging.

I am honouring the ancient practice of foraging in this chapter, sharing some popular, seasonal recipes for you to try. It's so much fun to do and is a great way to involve the whole family and explore nature. They also make some very thoughtful gifts and are always well received by loved ones.

Foraging tips
Before we dive into foraging, there are a few things to consider.

1 The first thing is to always positively identify your plant before harvesting or eating it. I will share an identification in these recipes, but it's really important to use a number of sources to identify your plant so that you are confident. I love to use the plant identifier app on my phone and a foraging book to help me if it is a plant that I have never foraged before. It's important to do your research and be safe.

2 Make sure you harvest sustainably. A foraging rule is to only take what you need and leave enough for the plant to regenerate and for wildlife to enjoy. It's important to avoid over harvesting, and make sure to forage your goodies from clean, uncontaminated areas.

3. Mind the rules. Make sure that you check local regulations before you forage as some areas like parks or private lands may have restrictions.

Sticking to these three practices, you can forage responsibly while staying safe and preserve the nature that you live in.

Wild garlic season

For me, the start of the wild garlic season in February feels like the start of a new year. Their shoots appear and bring an unmissable fragrance. We have an area near where I live that is full of wild garlic and I love to walk through it and breathe in the air. You can use wild garlic in lots of different recipes – the leaves and flowers are both edible.

> *Tips for foraging wild garlic:*
> - Be careful that you have identified wild garlic and not lily of the valley plants. Here is an illustration to show the difference. If you rub the leaf of the wild garlic, you can also identify what it is straight away as it has that distinctive garlicky smell.
> - Wild garlic flowers are an early nectar source for butterflies and bees, so take only as much as you need. Always take your time and pick leaves individually and never take the root of the garlic so that it can grow back each year.

Wild garlic

Lily of the valley

. . . LITTLE LUXURIES

Wild garlic pesto

> 100–150g wild garlic, washed
> 1 tablespoon lemon zest
> Juice of ½ lemon
> 50g sunflower seeds, toasted
> 60ml olive oil
> 2 tablespoons Parmesan or nutritional yeast

1. Put the wild garlic, lemon zest, lemon juice, toasted sunflower seeds and olive oil into a blender and blitz until combined.
2. When the pesto has reached a smooth consistency, add the Parmesan and blend until combined (you can also add nutritional yeast at this stage, but it must not be blended as it will change the consistency).
3. Add more olive oil if you would like a runnier pesto.
4. Transfer to a clean jar and top with a little more oil. This will keep in the fridge for up to 2 weeks.

I use toasted sunflower seeds in this recipe as they are cheaper than pine nuts and actually more sustainable to grow, but you could substitute these for pine nuts or even cashew nuts if you wish.

This pesto makes a great sandwich starter, can be drizzled on soup and stirred into pasta – it's such a low-waste option. If you don't have wild garlic to forage, you could try carrot tops, basil or rocket instead.

... LITTLE LUXURIES

Wild garlic salt

This wild garlic salt is a great way to capture the fresh garlicky flavour of this early spring delicacy, and once made, it can be stored for several months. Use as a finishing garnish on a whole host of dishes, from dips and freshly baked breads to platters of roasted seasonal veg.

> 50g wild garlic, washed
> 400g flaked sea salt

1. Place the wild garlic into a blender with 100g of the sea salt and blitz to form a thick, paste-like mixture.
2. Add the remaining salt to a bowl and combine with the wild garlic mixture.
3. Spread the mixture in a thin layer on a baking tray and leave to thoroughly dry out. This may take 1–2 days.
4. Once the mixture is completely dried out, transfer the salt to clean, sterile, reusable jars.
5. Label and store in a cool, dry place for up to 3 months.

This salt can be used in place of regular salt to season all of your dishes and to add to recipes. It is a lovely way to bring wild garlic into your recipes and celebrate the wild garlic season for a little longer. If you don't have wild garlic available, you can also make this recipe with lots of herbs. Oregano and rosemary would work well.

Elderflower cordial

This is such a delicious summertime cordial. I enjoy it diluted with sparkling water, but you can also add it to sparkling wine for a fresh cocktail.

Makes 1 litre
18 elderflower heads
1 litre water
600g granulated sugar
30g citric acid
1½ lemons

1. Collect the elderflower heads from a clean and pesticide-free area and ensure they are in full bloom and fragrant. Forage sensibly and leave lots for the pollinators.

2. Gently shake them to rid them of any insects or debris. Do not rinse the elderflowers as this will wash away the natural yeast of the elderflower. Place the elderflower heads in a large bowl.

3. In a large pan, add the water and sugar and bring to the boil.

4 Take the pan off the heat and add the citric acid, stirring thoroughly. Allow the mixture to cool a little, then pour it over the elderflowers and stir again.

5 Cover the bowl with a tea towel and set aside to steep for 24–48 hours.

6 When rested, strain the liquid through a muslin cloth or clean cotton tea towel. Pour into a clean, sterile, reusable bottle. Enjoy diluted 1 part cordial to 5 parts spritz of your choice.

The cordial will keep for up to 3 weeks. If you have not finished your cordial in this time, you can freeze it in ice-cube trays to add to your drinks later.

Elderflower champagne

Elderflower champagne is a delightful, lightly fermented drink made from elderflower blossoms you can make at home.

> 6 elderflower heads
> 2 litres water
> 300g caster sugar
> Juice of 1 lemon (keep the squeezed-out halves)
> 1 tablespoon apple cider vinegar

1 Gather the elderflower heads from a clean and pesticide-free area and ensure they are in full bloom and fragrant. Forage sensibly and leave lots for the pollinators.

2 Gently shake them to rid them of any insects or debris. Do not rinse the elderflowers as this will wash away the natural yeast of the elderflower. This is crucial for fermentation in

the recipe. Trim away as much of the stalk as possible, leaving mainly the flowers.

3 Bring the water to the boil in a large pan, then remove from the heat. Add the caster sugar to the water and stir until dissolved. Add the elderflowers, lemon juice and squeezed-out halves, and the apple cider vinegar to the sugar water and stir to combine.

4 Cover the pan with a clean towel and let it sit at room temperature for 48 hours, stirring the mixture once or twice each day. The natural yeast from the elderflowers should start the fermentation process and you will start to see bubbles on the surface of the mixture.

5 Leave the mixture to continue fermenting for another 24 hours, then strain the mixture using a muslin cloth or clean cotton tea towel.

6 Pour the strained liquid into clean, sterilised, reusable bottles. You will need to use very strong bottles for this part of the recipe, and I would recommend either swing-top glass bottles that have been designed for fermentation or to upcycle two plastic soda bottles for the task. Make sure you leave 3–5cm of headspace from the top of the bottles.

7 Seal the bottles and leave to ferment further at room temperature for 2 weeks. As the fermentation occurs you will notice that pressure builds up in the bottles and the bottles will look like they are beginning to expand. When this occurs, you will need to 'burp' the bottles by opening the tops slightly and releasing the pressure, before quickly tightening again.

8 After 2 weeks, refrigerate the bottles to slow down the fermentation process. If you aren't drinking it straight away, make sure you continue the process of burping the champagne to avoid pressure build-up.

9 Serve chilled – ideally on a sunny day!

Dandelion honey

Dandelion honey is a lovely vegan alternative to honey that contains all of the flavours of spring. It's really easy to make at home and enjoy all of the beauty of the humble dandelion.

Dandelions are one of the most common plants and often cursed in the garden as a perennial weed, but they are so very useful and shouldn't be overlooked. They provide an early food source for bees and other pollinators and are highly valued in herbal medicine and culinary practices for their wide range of health benefits and uses.

There are a number of dandelion recipes that I could share with you, but my favourite has to be dandelion honey, which is a vegan alternative to honey and tastes so floral.

> 100g dandelion petals
> 350ml water
> 1 lemon, sliced
> 300g sugar

1 Gather the dandelion heads from a clean, pesticide-free area.

2 Remove the yellow petals from the base of the flowers. It's a little time-consuming but will make a beautiful, flavourful honey.

3. In a large pan, combine the dandelion petals, water and lemon slices.

4. Bring the mixture to the boil and simmer for about 15 minutes.

5. Remove the mixture from the heat and allow to cool. Cover the pan with a clean cloth and leave it to steep overnight to get all of the flavour from the petals.

6. Strain the liquid through a fine-mesh sieve or muslin cloth.

7. Add the sugar to the strained liquid and return to a medium heat until the mixture reaches boiling point.

8. Reduce the heat and simmer, stirring occasionally until the liquid thickens to a syrup consistency.

9. To check if the honey is ready, place a teaspoon of the honey on a cold plate and allow it to cool. If it thickens to your liking, it's ready! If not, continue simmering and testing until you achieve the consistency you want.

10. Take the mixture off the heat and allow to cool. Pour into a clean, sterile, reusable jar.

11. Label your dandelion honey and store in a cool, dark place. It can be kept in the fridge for up to 3 months.

Your dandelion honey can now be used to sweeten pretty much anything you choose! I love it drizzled over fresh fruit and pancakes; it is such a lovely way to make use of the dandelion – a beautiful wildflower that is often overlooked!

Pickled magnolia

Another welcome signal of spring is the arrival of magnolia, and it is one of my favourite flowers. Magnolia blossoms are typically showy, with smooth petals that can range from creamy white to shades of pink and purple.

Pickling magnolia buds is a traditional practice, especially in East Asian cuisines; they taste similar to pickled ginger. You can use them on salads, as a garnish, or if you are like me just eat them straight from the jar.

> 8 fresh magnolia buds
> 150ml white vinegar
> 150ml water
> 1 tablespoon granulated sugar
> 2 teaspoons sea salt

1 Pick the magnolia buds while they are young and tender, while the flower is still closed.

2 Gently rinse the magnolia buds to remove any debris.

3 Make a brine solution by combining the vinegar, water, sugar and salt in a pan and bring to the boil, stirring until completely dissolved.

4 Place the magnolia buds in a clean, sterile, reusable jar, then pour the hot pickling brine over the magnolia buds, making sure they're completely submerged.

5 Once cooled, place the jar of magnolias in the fridge and allow to pickle for a few days.

6 Store in the fridge for up to 3 weeks.

Rose water

This is a recipe that you can use in your food, drinks and even skincare. I like to collect the flowers from the rose bushes in my garden to make it.

Rose water has strong anti-inflammatory properties and soothes irritations and redness in your skin. It's commonly used as a toner in skincare, and I am going to show you how you can make your own.

> 100g rose petals
> 2 litres water

1 Collect the petals from roses that are edible and free from pesticides.

2 Make sure the petals are free from dust or insects and rinse them if necessary.

3 Add the petals (you can also use dried petals) to a pan and add enough water to cover them (about 1 litre).

4 Place a small heatproof bowl inside the pan (this will collect the rose water).

5   Now place the lid on the pan upside down. You need to make sure there is a tiny gap between the dish and lid so that the rose water condenses on the lid and can be caught in the bowl.

6   Bring to a simmer and add ice cubes to the lid. When the rose water steam hits the lid of the pan it will then condense and be caught in the bowl.

7   You'll need to add more ice cubes as it's simmering – the whole process should take about 30 minutes.

8   Allow the rose water to cool before transferring to a clean, sterile, reusable container.

9   This will store for up to 3 days in the fridge. If you would like it to last longer, you could freeze it into ice cubes to defrost at a later date when needed.

Elderberry syrup

Making elderberry syrup is the best way to celebrate the change in seasons and welcome autumn. You will find that the black, shiny elderberries begin to appear from late August. I forage the elderberries and make this delicious syrup to help boost my immunity over the winter months.

When foraging elderberries, make sure that they are ripe; it is quite easy to see as the unripe berries are hard and a greenish red colour. You need to make sure you wash your elderberries and remove them from the stalk using a fork. This is an important step as the stalk is poisonous and may cause an upset stomach.

Here is the recipe that I use to make the syrup. It will store in the fridge for up to 3 months and you can also freeze a batch to bring out in January. As well as a way to bring in the seasons, connect with nature and boost your wellbeing – this recipe will also save you money. Elderberry syrup is a popular way to ward off the common cold but is expensive to buy in the shops – an extra little bonus for you! I like to take a tablespoon every day – you can also add a tablespoon to porridge, granola, smoothies or your favourite tea. I also like to make wellbeing gummies with this recipe – see page 160.

> 500g fresh elderberries
> 5 slices of fresh ginger
> 1 cinnamon stick
> 500g caster sugar
> Juice of 1 large lemon

1. Rinse the berries and remove them from the stalk using a fork (discard any green berries as you remove them).

2. Place the berries in a pan and add just enough water to cover them.

3. Add the ginger and cinnamon stick and bring to almost boiling, then simmer for 20 minutes until soft.

4 Remove the cinnamon stick and mash the mixture.

5 Strain the berries through a fine-mesh sieve or sieve lined with a muslin cloth.

6 Return to the pan and add the caster sugar and the lemon juice.

7 Simmer for a further 10 minutes until the sugar is dissolved, then remove from the heat and allow to cool.

8 When cooled transfer to a clean, sterile, reusable bottle.

Sloe gin

In late September or early October, I love to welcome winter by making sloe gin. It's such a simple recipe and makes a great gift for Christmas. I like to make my favourite sloe gin fizz at Christmas using this recipe.

> *Makes 2 large jars*
> 750g sloes
> 250g granulated sugar
> 750ml gin

1 Gather, wash and dry the sloe berries. Some people advise to pick sloe berries after the first frost, but I have had better results by picking them when they appear ripe on the tree. If I wait for the first frost, I have found that they go too squishy and are often half eaten. To mimic the first frost, I have a handy little tip. When you have gathered and washed the sloes, pop them in the freezer overnight.

2. Divide the sugar equally between two sterilised large jars or wide-mouthed bottles.

3. Divide the frozen berries equally between the jars then divide the gin equally between the jars as well.

4. The gin should cover the sloe berries; if it doesn't, add a little more gin to make sure the berries are completely submerged.

5. Pop the mason jars in a dark place and turn them upside down each day until you notice that the sugar has completely dissolved. Then you can leave them to turn into a lovely sloe gin.

6. Check on them every so often and give them a gentle shake. You will notice that the colour changes from a beautiful deep pink to purple.

7. In December, your sloe gin will be ready to enjoy!

If you are gifting this lovely drink, find a nice bottle and make sure that it is clean and sterile. Strain your mixture using a fine-mesh sieve or muslin cloth (be aware that this will stain anything that it touches). Transfer to the bottle and label.

Rosehip syrup

Rosehips are an unsung hero in my opinion, I love to use rosehip oil to impart its beneficial properties and brighten my skin, and I also use them to make a beautiful, vitamin-rich cordial to enjoy in the summer months.

> 500g rosehips
> 750ml water
> Juice of 1 large lemon
> 200g granulated sugar

1. Gather and wash the rosehips, add to a blender and briefly blitz to roughly chop them.

2. Add the rosehips to a large pan and cover with the water. Bring to the boil then lower the heat and simmer gently until the rosehips become soft. This should take approximately 15 minutes.

3. Remove from the heat and allow the mixture to cool for at least 15 minutes.

4. When cooled, pour the mixture through a fine-mesh sieve or sieve lined with a muslin cloth and allow the liquid to drain through, this will take a little time – be patient!

5. When you have your rosehip juice, add it to a pan with the lemon juice and sugar and stir over a low heat until the sugar has dissolved completely into the liquid.

6. Bring to a rapid boil for a few minutes to reduce the liquid and make a syrup.

7. Allow to cool and then pour the rosehip syrup into a clean, sterilised, reusable bottle.

8. This should last in the fridge for up to 3 months.

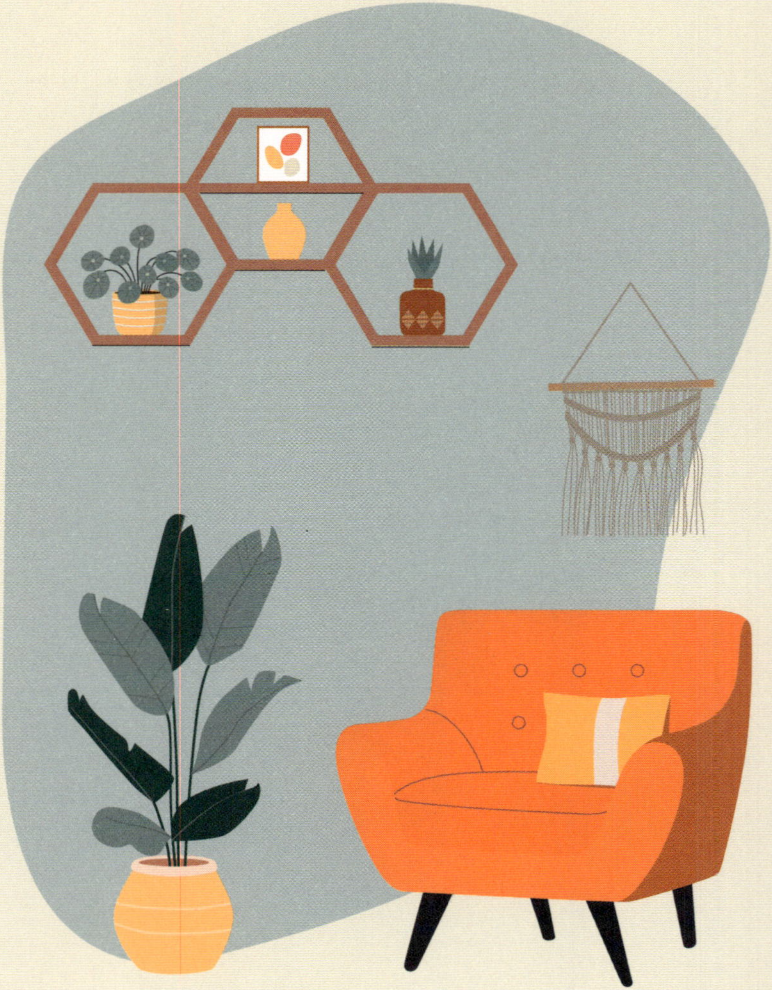

MY PLASTIC-FREE DECOR

MY PLASTIC-FREE ...

LOW-WASTE AND SEASONAL DECORATIONS

I absolutely love making and creating seasonal decorations for my home. I find it such a simple and mindful way to connect with nature, avoid all of the shop-bought decorations and make my home look beautiful.

Here are some of my favourite ways to decorate my home. I have selected really easy crafts that you can replicate either on your own, or with friends and family.

Seasonal wall hanging

I love to decorate my home with seasonal wall hangings. You will find that I usually have at least two or three around the house at one time because they make me so happy. It's like decorating your house with flowers, but they last so much longer and create such a welcome feature of the season, which I find a wonderful way to celebrate the different times of the year. You can customise them to reflect your personal style and the specific season you're highlighting.

> Botanicals
> Needle and cotton thread
> A branch to display your wall hanging from
> Twine

1 Prepare the botanicals that you would like to hang from your wall hanging. I tend to choose colours and decorations that represent the season that I am celebrating. Some of my favourite decorations to use are eucalyptus, dried orange slices, lavender, hydrangeas and roses.

2 Thread or tie the cotton around the decorations and then attach to your branch. I like to make bunches of lavender to hang but tie the eucalyptus and roses individually.

3 Trim any remaining threads and hang your decoration using the twine.

Festive wall decoration

This is a version of a wall decoration that I make every Christmas.

> 'Clay' stars (see page 192)
> Paper stars
> Dried orange decorations (see page 195)
> Pine cones
> Brass bells
> Pine sprigs
> Needle and cotton thread
> A branch to display your wall hanging from
> Twine

1 Prepare the decorations that you would like to display on your wall hanging.

2 Thread or tie the cotton thread around the decorations that you would like to hang and then attach these to your branch. I tend to tie white cotton around the decorations as my walls in the house are light and they give the illusion that the decorations are floating.

3 Trim any remaining threads and hang your decoration using the twine.

Wax leaves

Preserving autumn leaves in beeswax is a lovely way to capture the beauty of the changing season and make decorative pieces that can last through the whole of winter. It is a seasonal activity that I always look forward to; as soon as the trees fill with all the autumnal colours and the leaves start to fall, I gather them up with my daughter and we bring them home to dip in beeswax.

> Freshly fallen autumn leaves of vibrant colours
> Soy wax or beeswax

1 Melt the wax of your choice using the double boiler method. When the wax is liquid, dip one leaf at a time into the wax, holding on to the stem. Submerge it fully into the liquid wax until completely covered, taking care as the wax will be hot.

2 Lift the leaf out of the wax and allow the excess to drip back into the liquid wax.

3 Wait for a few seconds for the wax to solidify and then place the leaf on a non-stick surface.

4 Display the leaves around your home in any way you like. I usually make simple garlands with them and hang them up around our home, or make them into a wall hanging, arranging them from yellow through to brown like a wave of autumnal colour.

The wax helps to preserve the colour in the leaves and celebrates the beauty of autumn in your home decor. Once they have been preserved in wax these leaves will last for up to a year, where their colour will gently change over time.

'Clay' stars

Another tradition I love is making these homemade star garlands. They are made from just two store cupboard ingredients, are so simple to create and look just like white clay. The decorations last for a number of years and you can make them in any shape you like. Our favourite in this house are stars but we have also made ghosts for Halloween.

> 125g cornflour/cornstarch
> 250g bicarbonate of soda
> 375ml water
> Cotton thread or twine for hanging

1. Preheat the oven to 140°C/120°C fan/gas mark 1.
2. Put the cornflour, bicarbonate of soda and water into a pan and mix together.
3. Place the pan over a medium heat until it becomes the consistency of mashed potatoes. Remove from the heat and leave to cool.
4. Dust your work surface with cornflour and roll the clay out to a thickness of 5mm.
5. Using biscuit cutters, cut out your shapes in whatever shape you like. At this point you will also need to make a hole in the shapes so that you can hang them when they have dried. I like to use a little skewer for this.

6 Bake the clay shapes in the oven for 10 minutes, then turn them over and bake for a further 10 minutes.

7 Alternatively, if you have an air fryer with a dehydrate option, or dehydrator, dehydrate until they are fully dry.

If you use large shapes for your decorations, you may need to leave them in the oven for a little longer. When they are ready, they will be firm, dry and white in colour. You can also file and sand any cracks that have appeared in the decorations when baking, but I must admit I like them to look a little rustic.

MY PLASTIC-FREE ...

Dried orange decorations

I have been making these dried-orange decorations for years and I absolutely love them. They smell amazing while you're making them, the smell lingers on wherever you hang them, and they look absolutely beautiful. I make them as a tradition each year, they do keep well if you would like to keep them for longer, and they are completely compostable too.

> 3 large oranges
> Cotton thread or twine for hanging

1 Slice the oranges as thinly as possible using a sharp knife.

2 Use a cloth to soak up the moisture of the oranges, this will speed up the process of drying them.

3 You can now dehydrate the orange slices until they have completely dried, or you can dry them in the oven. If using the oven method, preheat the oven to 110°C/90°C fan/gas mark ½. Bake for 20 minutes then reduce the oven temperature to 90°C/70°C fan/gas mark ¼ and bake for a further 1 hour 20 minutes, turning them intermittently.

4 Start checking on the oranges and you will notice that the smaller slices will dry quickly so take these out of the oven as they dry. I tend to take the oranges out of the oven or dehydrator before they are completely dry as they will carry on drying when hung. I find it's always better to be under rather than over on your drying time. A good indicator is to take them out when you can see the light shining through the orange.

Popcorn garland

Popcorn garlands are really simple to make and are such a charming traditional decoration for festive occasions. I love to hang them on my tree as a compostable alternative to tinsel and I think they look so effective.

> A handful of popcorn kernels
> Needle and cotton thread

1 Pop the popcorn kernels on the stove or in the microwave and allow to cool completely.

2 You can leave the popcorn for a day so that it becomes slightly stale, as fresh popcorn can be more difficult to work with.

3 Cut a length of thread to the length that you would like your garland to be.

4 Thread a needle and tie a bead or button at the end of the thread to keep the popcorn from slipping off.

5 Gently push the needle through the centre of the popcorn, and continue adding popcorn until you have made your garland.

6 You can include dried accompaniments to your popcorn garland, like dried orange slices, or you can leave them as they are.

Carefully drape the garland on your Christmas tree or around your windows and door frames. Popcorn garlands are completely compostable, but do not store well. They are more of an annual tradition in our home, which we use to decorate the tree and then compost in the new year.

Upcycled paper stars

I am a fond reader and have so many old books in my house, some where the spine has come away from the pages, others that are too tattered to donate to the charity shop. I love to turn these pages into beautiful things to decorate my home – I hang them up throughout the year, but I find they are particularly popular with guests during the festive season. You can also use music paper and old maps too. Each star is just two squares of paper: the larger the square, the bigger the star.

> Old book pages, sheet music or maps that need recycling
> Scissors
> Paper tape
> Glue
> Piece of thread or string

1 If your piece of paper isn't square already, fold in two of the corners diagonally and then cut off the spare rectangle at the end, so you have a square shape to work with. If your paper is square already, make the folds anyway, as you will need them to make your star.

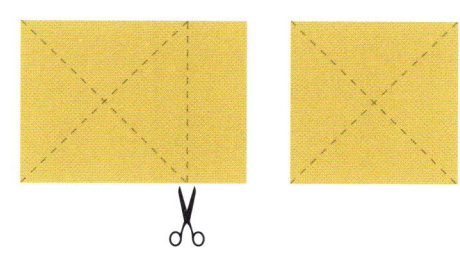

2 Then fold over the paper edge to edge in both directions, so it now has four fold lines running across it.

3 Using the scissors, cut almost halfway along the folds you made in step 2 (but don't cut along the diagonal folds of step 1).

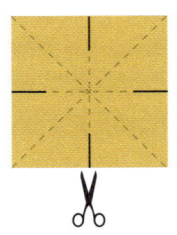

4 At each cut, fold the sides in to meet the diagonal line.

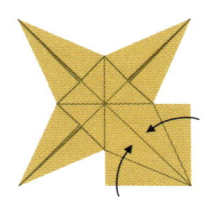

5 Now apply glue to one of the folded-over triangular flaps, and bring the other triangular flap across to completely cover it.

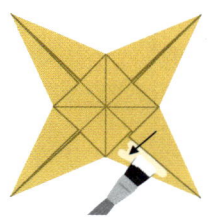

6 Repeat this process until you have a four-pointed star.

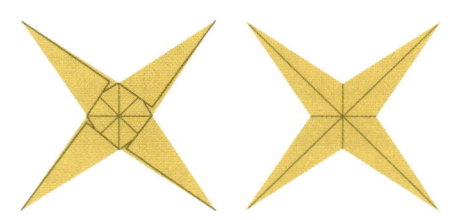

7 Now take a new piece of paper and follow steps 1–6 to create a second four-pointed star.

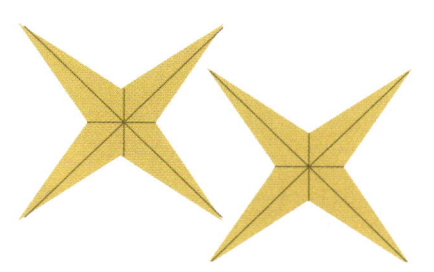

... DECOR

8 Use a small piece of sticky tape to fix a piece of thread or string inside one of the stars.

9 Now apply dots of glue to both stars at the places where they will touch. Press the two halves of the completed star together, and when the glue is dry hang it up.

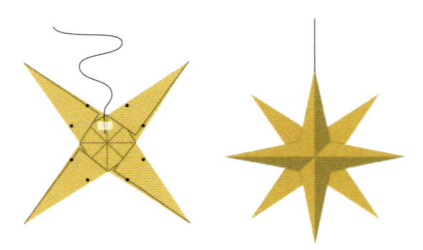

Enjoy your stars! I love to decorate my trees, hang them from garlands or simply place them around my house to add a little festive cheer.

... DECOR

Paper Christmas trees

I love to make Christmas trees out of old vintage books, maps, music paper or even simple brown paper. They are so charming, and last for years. They are incredibly simple to make, so you may find yourself becoming a little obsessed and making a little forest as I seem to do.

> Pages of an old book, sheet music or map that needs recycling, cut into squares
> Scissors

1 Prepare a square piece of paper. If this is a piece of sheet music or the page of a book, a simple way to do this is to fold the top edge to the side edge to get a perfect diagonal. Trim the bottom edge so that you are left with only the triangle. Now when you unfold your triangle you will have a square shape.

2 Open the square and fold diagonally in the opposite direction, so that you now have a cross from corner to corner. Now flip the paper to the other side and fold it in half horizontally and vertically so that it also has a straight cross through the square.

3 Now push the diagonal creases inwards and draw the straight creases outwards. This should fold down into a small square shape with two square sides on the outside and the diagonal edges gathered inside.

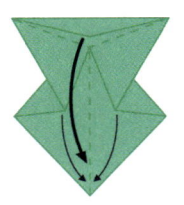

4 Take the front square and fold the edge into the middle to make a small triangle. This will leave a crease line for you to follow.

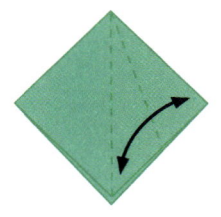

5 Unfold this fold that you created and flatten it into the middle line so that it now lies symmetrically on the page. Flip your square over and do the same with the opposite side of your page. Carry on until you have made this pattern with each edge.

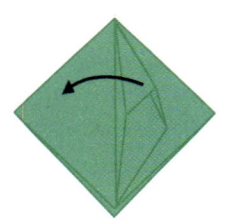

6 Now use a pair of scissors to cut the bottom of the tree so that it makes a flat edge.

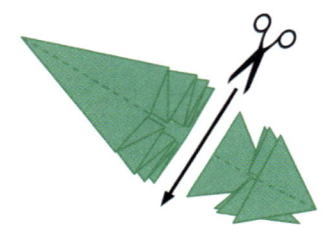

7 Lie your shape flat so that it is a triangle and use a pair of scissors to cut three or four slits on the sides.

8 Fold the slits that you have made to 45 degrees inwards and flip each layer, then repeat the process on each page of your tree.

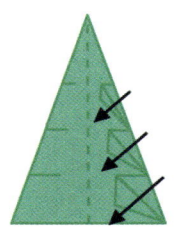

9 Now stand your Christmas tree up and use to decorate your home.

There seems like quite a few steps to these decorations, but they are truly simple to make. I like to put my favourite cosy box set on the television, pour a cup of something warm and get creative in the evening. My daughter likes to join me in the production line too!

. . . DECOR

MY PLASTIC-FREE ...

Recycled wrapping paper

Another thing that I love to do is wrap small gifts in book pages. It is my absolute favourite way to wrap gifts, and it has that perfect blend of luxury, vintage and sustainability.

I also like to wrap gifts in newspaper and old maps. For a few months before Christmas I ask my mum to start saving up her newspapers and use these to wrap gifts. It's an old myth that newspaper inks are toxic. Modern newspaper inks are perfectly safe and can also be composted after use. Brown paper is also a lovely way to wrap gifts and completely compostable too.

Choose wrapping paper and cards made from natural materials

I love to make my own, but if you find yourself needing to buy wrapping paper, there are a couple of things that you can do to make sure that it is plastic-free, recyclable or compostable.

Look for wrapping paper and cards that are made from paper only. Anything that has glitter or foil as part of the pattern will not be able to be recycled, and glitter is another form of microplastic that is so terrible for the environment. Opting for paper and card without these things is a good option. If you're wanting to test the wrapping paper, you can always do the 'scrunch test'. Scrunch a piece of wrapping paper in your hand and when you open your fist if it stays scrunched together then it is made of paper, but if it opens up quite quickly then it will have a coating on it that isn't paper and will not be recyclable.

Making special occasions more sustainable
I absolutely love celebrating special family occasions, particularly my children's birthdays, and there are a few very simple ways that I do this to keep them as low-waste as possible.

Opt for digital invitations instead of paper
A really simple way to begin your special occasions with less waste is to send out digital invitations instead of paper. They're really simple to make, and there are so many applications and templates that you can use to help you.

I really enjoy using Canva (a free design app) as it is great for beginners, and you can just type in the type of invitation that you'd like, and it basically creates the whole thing for you. Not only are digital invitations waste-free but they're also cost-effective and convenient for both you and the recipient.

Choose reusable decorations made from natural materials or make your own
I love to decorate for parties, it's also one of my favourite things to do for the children's birthdays; I love to see the look on their faces when they come downstairs to a decorated room in the morning. The problem is that lots of traditional party decorations are full of plastic and are single-use. To avoid this, I just stick to a few simple rules:

1 **Choose reusable decorations**. I have bunting for both of my children that I reuse year after year, and pom-poms and honeycomb balls that I bring out every year instead of

balloons. I really feel that they are just as effective as balloons and come in so many bright colours.

2. **Choose natural materials.** I both make and buy paper decorations, as when they have come to the end of their timeline, they can be composted. I also use flowers from my garden, or a local florist, so that I can decorate table settings and brighten up spaces with them. Again, because they come in so many different colours and varieties you can personalise them to any party. You could even let your guests take them away at the end of the day as an extra gift!

3. **Choose reusable or compostable plates and cutlery.** At the last party I threw I purchased some bamboo plates that are reusable, and each year I add to my little kit of reusables so that I'm not just throwing away items year after year.

4. **Borrow what you can.** I was able to borrow tablecloths, tables and chairs so that I didn't have to buy them. Borrowing and sharing is such an effective way of lowering your environmental impact in so many situations.

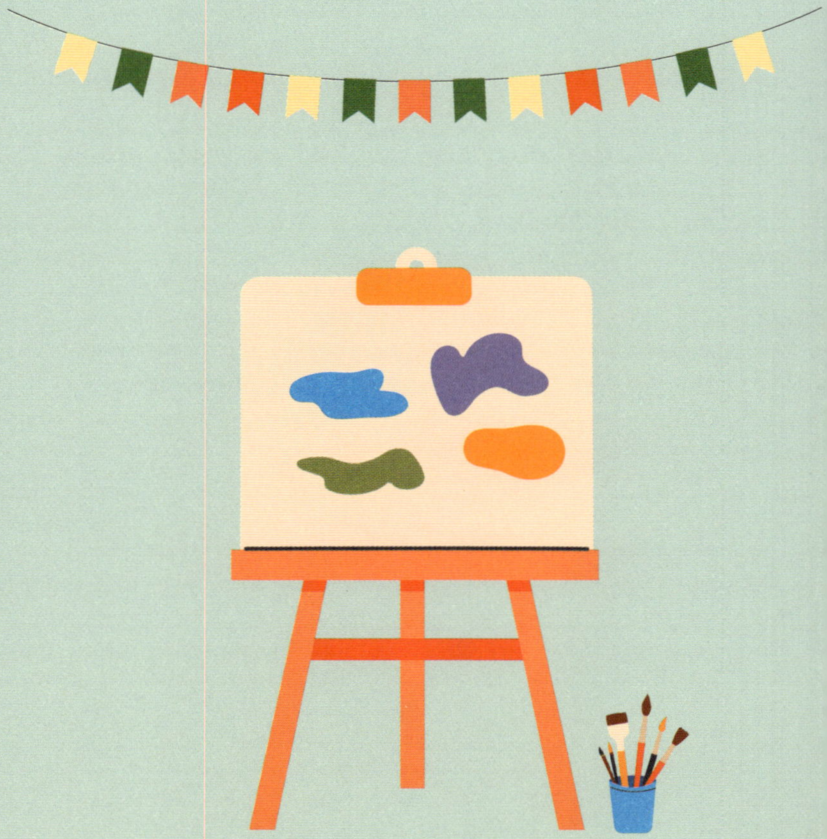

MY PLASTIC-FREE FAMILY FUN

LOW-WASTE CHILD'S PLAY

I have two very young children who absolutely adore to craft and play, and would happily have me making up potions and activities for them all day long.

Here are a few of our favourite activities that we enjoy doing together and keep them captivated for hours! These are our favourite ways to play, keeping them entertained so much more than plastic toys do.

Cloud dough

Making cloud dough with conditioner is a fun and easy variation of playdough that results in a smooth, silky texture. I absolutely love to make this dough as it's gently scented and smooth. I have found that other homemade doughs can be quite drying on my children's hands, which makes this a perfect alternative. You can either use food colouring extracted from food, or natural food colouring to make this fun activity.

> 250g cornflour/cornstarch
> 125g hair conditioner
> Natural food colouring, powders or botanicals

1. Mix the cornflour and conditioner until it starts to form a dough.
2. Once the mixture starts to resemble dough, knead it until it becomes smooth and pliable. If the dough is sticky, add a little more cornflour, if it is too dry add a little more conditioner.
3. Separate into as many portions as you would like to make.

4 Add a drop of food colouring or powders, like beetroot and spirulina, to each piece in the colour that you would like. At this point you could also add botanicals like lavender buds, rose petals or calendula to add interest and fragrance too.

5 You're now ready to play with the dough!

When you have finished playing with the dough, you can keep it in an airtight container for a few weeks.

Handmade waste-saving giant crayons

If you have young children, then I'm guessing that you have a lot of broken crayons in the house! Making these giant crayons is a great way to use up all of those crayon pieces.

1 Preheat the oven to 120°C/100°C fan/gas mark ½.

2 Peel the covers from the crayon pieces and cut them into small pieces (no bigger than 1cm).

3 Place them into a heatproof mould. I have a muffin tray that I picked up from a charity shop for this purpose.

4 Melt the crayons in the oven for approximately 10 minutes, you will see that they have become a liquid consistency.

5 Carefully remove the mould from the oven and allow the melted crayons to cool and solidify. You can place in the fridge to speed up the process, or you can just leave them to cool at room temperature for a little longer.

6 When hard, remove from the mould. Your crayons are ready to use, have fun!

I have found that these crayons are so good for little hands, and kids love using them because they are much more interesting than one colour on its own. I tend to group colours together to create the crayons so that they still have a little order, but you can go wild if you like and just throw them all in.

DIY paint

Making paint with cornflour is so much fun and such an easy project. It is especially suitable for those looking for non-toxic, homemade paint options. As well as using them for painting I also like to use them as an outdoor activity by letting the children's imagination run wild painting the paving stones. Then I just wash it away when we are finished.

> 250g cornflour/cornstarch
> 250ml water
> Natural food colouring

1 In a bowl or pan, mix the cornflour with the water until smooth and well combined.

2 You can adjust the consistency by adding more cornflour for a thicker paint or more water for a thinner paint.

3 Divide the mixture into separate containers for each colour of paint you would like to make. I use a muffin tin for this.

4 Add natural food colouring to each container. You will only need one or two drops for this, stir well until the colour is evenly distributed. You can adjust the colour intensity by adding more drops, or mix food colouring together for different colours.

In case you need a little recap on colours, here is what different combinations will make:

- Red and blue = purple
- Red and yellow = orange
- Blue and yellow = green
- White and black = grey
- White and red = pink
- Red, yellow and blue = brown

You can store your paint in airtight jars for up to a week.

Seasonal nature board

This is by far my children's favourite outdoor activity, and they are always asking to take one out with them when we go for a walk. You can make them in all different shapes and sizes; a particular favourite in our house is a lion's face which we surround with flowers for a mane. All you need is a piece of cardboard and some pens . . . and a day out in nature!

> Cardboard
> Pens
> Utensil for making holes (e.g. skewer)

1. Take a piece of cardboard and cut out a shape you would like to make. I find circles are super-easy and work well as you can draw two eyes and a mouth in the middle.
2. Using your skewer (or similar tool) make holes around the edge of the circle.

3 Take a trip out in nature and collect what you need to pop through the holes. I find that daisies and dandelions work great for this activity, as do small leaves and wildflowers – but please make sure to leave enough behind for the pollinators!

I like to reuse these through the summer, you can decorate them with lots of different things – leaves and grasses work particularly well too!

Nature colour sorter

Creating a nature-inspired colour sorter involves using natural materials to categorise and sort items based on their colours. This is such a playful and educational seasonal activity, and one that my children still enjoy. The process of making the colour sorter, going on a walk in nature, gathering the little bits of nature and talking about all of the different steps, are all fun little activities to engage in.

> Egg carton to sort your natural treasure
> Paints to decorate your colour sorter

1 Remove the lid of the egg carton.
2 Paint colours inside each section of the tray to match with the treasures of nature that you would like to collect.
3 Plan your day in nature to gather your treasures.

Making a nature colour sorter is a great way to encourage creativity with your little ones and also helps to promote learning and exploration of the natural world. Be careful to supervise your children during this activity as not

. . . FAMILY FUN

all of nature's treasures are edible! Here are a few ideas of things that we like to collect:

- Green for leaves and grasses
- Brown for pine cones, twigs and soil
- Pink for rose petals and cherry blossom
- Blue for berries and cornflowers
- Purple for pansies, petals and blackberries
- Yellow for daffodils, dandelions and daisies

Nature loom

Making a nature loom is a creative and hands-on way to engage with natural materials and create beautiful woven art. They are so simple to make and can be placed around your home as decoration too.

> Sturdy branch
> Yarn or twine
> Natural materials to weave into the loom

1. Go for a walk in nature and find a sturdy branch for your loom – a Y-shaped branch is absolute perfection! Or alternatively you can gather three or four branches to overlap one another and tie together.
2. Tie one end of twine to one end of the branch and stretch the twine to the opposite end, ensuring that it stays taught.
3. Keep weaving the twine from side to side, keeping it taught, and once you reach the end of your branch, tie and secure the twine.

4. The spacing of the twine is up to you, depending on how dense you would like your weave to be.
5. Now it's the fun part! Gather items that you would like to add to the loom. Leaves, flowers with long stems and long grasses work well for this activity.
6. Using the items that you have gathered, weave them over and under the threads that you have attached.

This is such a relaxing activity and, again, I love the process of making the loom, and hunting for treasures to display on it. It can be done throughout the changing seasons and you could plan a lovely, relaxing day around it.

Nature mandala

This is a similar activity to the loom, but encourages lots of creative play with the different little objects. I love that the circular design of the mandala symbolises how interconnected we all are; it is a beautiful symbol and such a lovely way to explore the great outdoors. You can gather your items and bring them home or make one in a little spot while you are walking among nature for other people to enjoy.

> Materials collected from nature
> Twigs and sticks
> Grasses
> Flowers
> Stones

1. Take a walk in nature and gather the natural materials that you would like to use to make your mandala. I have included an illustration of a simple mandala that you may like to recreate.

2. When collecting your natural materials, think about collecting equal numbers of them, e.g. 8 pebbles, 12 petals etc.

3. When you have collected all of the materials, lay them out in any order that you would like. Take your time to make patterns and shapes in the mandala that you feel connected to.

4. Take the time to talk to your children and explore the various parts of the mandala, how the different elements feel, the assorted colours, etc.

Upcycling paper
Handmade recycled paper

This is such a lovely activity that I do with my children to recycle old craft paper, and they love to do this as a way to give papers a new life. I find that not only do my kids enjoy this craft, but it also indirectly teaches them about how valuable resources can be, and that there are ways to look at how things can be reused without simply putting them in the recycling.

> Scrap paper
> Blender
> Large bowl or container
> Stick/immersion blender
> Fine-mesh sieve or screen
> Cloth towels
> Sponge

1. Start by shredding your scrap paper into small pieces. The smaller the pieces, the easier you will find they blend.

2. Place the shredded paper into the blender and cover with water. Blitz until the mixture becomes a pulp.

3. Place the pulp into a large bowl or container and cover it with water. Let it soak for several hours. This helps to break down the fibres and prepare them for forming into new paper.

4. Use a stick blender to further break down the paper into a smooth pulp.

... FAMILY FUN

5 Fill your basin with water and add a small amount of the pulp to the water and stir to disperse it evenly through the water; the more pulp you add, the thicker the paper will be.

6 Place the fine-mesh sieve or screen over the pulp mixture and lift it out evenly, allowing the excess water to drain away.

7 Transfer the pulp on the screen onto a cloth by flipping it over. Use a sponge or cloth to gently press out excess water.

8 Peel the screen off the new paper that you have made and allow it to dry.

9 Repeat this process to make as many sheets as you like.

10 Once the sheets have begun to dry, air-dry them flat on a drying rack or hang them out to dry so that they are well ventilated.

Plantable seed paper

This is another simple upcycling craft that I love to do with my children. Again, I tend to use upcycled craft paper or writing paper but during the process I add seeds that I have saved from the garden and dried petals to the paper.

> Scrap paper
> Blender
> Large bowl or container
> Stick/immersion blender
> Seeds and/or dried flower petals
> Fine-mesh sieve or screen
> Cloth towels
> Sponge

1 Start by shredding your scrap paper into small pieces. The smaller the pieces, the easier you will find they blend.

2 Place the shredded paper into the blender and cover with water. Blitz until the mixture becomes a pulp.

3 Place the pulp into a large bowl or container and cover it with water. Let it soak for several hours or overnight. This helps to break down the fibres and prepare them for forming into new paper.

4 Use a stick blender to further break down the paper into a smooth pulp.

5. Fill your basin with water and add a small amount of the pulp to the water and stir to disperse it evenly through the water; the more pulp you add, the thicker the paper will be.

6. At this point, add any seeds or dried flower petals that you would like to be added to the paper.

7. Place the fine-mesh sieve or screen over the pulp and seed mixture and lift it out evenly, allowing the excess water to drain away.

8. Transfer the pulp on the screen onto a cloth by flipping it over. Use a sponge or cloth to gently press out excess water.

9. Peel the screen off the new paper that you have made and allow it to dry.

10. Repeat this process to make as many sheets as you like.

11. Once the sheets have begun to dry, air-dry them flat on a drying rack or hang them out to dry so that they are well ventilated.

I love to cut the seed paper into gift tags or small cards to give to loved ones. They make a great party favour gift too. Always be careful to use native seeds when making seed paper, and all you need to do is plant the paper in a pot of compost and watch the seeds grow.

Dyeing clothes using food waste

Did you know that food waste can be used to make natural clothes dyes? Although I am quite the amateur, I'm a little obsessed with using natural dyes to upcycle clothing. Here are four lovely natural dyes that are easy to make and beginner-friendly and so much fun to do as a family. If I can do it, you definitely can!

Avocado dye

Dyeing fabrics with avocado pits and skins brings out the most beautiful soft pinks and salmon hues that you will fall in love with instantly.

- Soya milk (this is optional, but it is a natural mordant and will make the dye more stable)

- Around 5 avocado pits and skins (the more avocados, the more intense the colour will be)

- Natural fabric that is light in colour (usually cotton, linen, silk or wool)

1. Prepare your fabric by washing it with a mild detergent – this allows the dye to adhere better to the fibres of the material.

2. If using soya milk as a mordant you will need to weigh the fabric to determine how much you need. Measure out the soya milk to weigh 10–15% of the fabric weight and dissolve in a large pan of water – just enough to submerge the fabric.

3. Bring the liquid to the boil and submerge the fabric into it. Simmer for about an hour and then rinse and set aside. This step is optional but will help the longevity of the dye.

4. Rinse the avocado pits and skins to remove any flesh. You can also use frozen avocado skins and pits (I freeze mine until I have enough to make the dye).

5. Place the avocado pits and skins in a large pan and cover with water.

6. Bring the mixture to the boil, then reduce and let it simmer for 1–2 hours. The longer you let it simmer the more concentrated your dye will become. You'll notice the water turning a brownish pink colour. Once the dye has reached an intensity that you like, remove the pan from the heat and allow it to cool a little.

7. Strain the liquid through a colander to remove the pits and skins.

8. Wet your pre-washed fabric thoroughly with water. This helps the dye to penetrate the fabric more evenly.

9. Submerge the wet fabric into the dye bath, ensuring it is fully covered and immersed in the liquid.

10. Bring the dye bath back to the heat and let the fabric simmer for about an hour, stirring occasionally to make sure the colour penetrates the material evenly.

11. Take the pan off the heat. Leave the fabric in the dye bath for as long as you wish until the fabric reaches a colour that you like.

12 Remove the fabric from the dye bath and rinse under cold water until the water runs clear.

13 Wash the dyed fabric with a mild detergent and hang to dry, out of direct sunlight, which can cause the colour to fade.

> *Tips:*
> The final colour of this dye will vary depending on the fabric type, number of avocado skins and pits used, and how long you dye the material for – enjoy the process of experimenting!

Onion skin dye

Onion skins create natural dyes that produce beautiful yellow, orange and brown hues. You can use either yellow or red onion skins – the yellow skins will produce a range of yellow to brown hues whereas the red skins will produce deeper, reddish-brown tones.

You will need to collect a large pot of the onion skins for this process; I usually just collect them in a bag until I have enough.

> Soya milk (this is optional, but it is a natural mordant and will make the dye more stable)
> Onion skins from yellow or red onions
> Natural fabric that is light in colour (cotton or linen are great options)

1. Prepare your fabric by washing it with a mild detergent – this allows the dye to adhere better to the fibres of the material.

2. If using soya milk you will need to weigh the fabric to determine how much you need. Measure out the soya milk to weigh 10–15% of the fabric weight and dissolve in a large pan of water – just enough to submerge the fabric.

3. Bring the liquid to the boil and submerge the fabric into it. Simmer for about an hour and then rinse and set aside. This step is optional but will help the longevity of the dye.

4. Rinse the onion skins to remove any debris. For a rich colour, you will need a large pan of onion skins.

5. Place the onion skins in a large pan and cover with water.

6. Bring the mixture to the boil, then reduce and let it simmer for 1–2 hours. The longer you let it simmer the more concentrated your dye will become. You'll notice the water turning a rich golden colour.

7. Once the dye has reached an intensity that you like, remove the pan from the heat and allow it to cool a little.

8. Strain the liquid through a colander to remove the skins.

9. Wet your pre-washed fabric thoroughly with water. This helps the dye to penetrate the fabric more evenly.

10. Submerge the wet fabric into the dye bath, ensuring it is fully covered and immersed in the liquid.

11. Bring the dye bath back to the heat and let the fabric simmer for about an hour, stirring occasionally to make sure the colour penetrates the material evenly.

12. Take the pan off the heat. Leave the fabric in the dye bath for as long as you wish until the fabric reaches a colour that you like.

13. Remove the fabric from the dye bath and rinse under cold water until the water runs clear.

14. Wash the dyed fabric with a mild detergent and hang to dry, out of direct sunlight, which can cause the colour to fade.

Tips:
The final colour of this dye will vary depending on the fabric type, amount of onion skins used and how long you dye the material for – experiment to see what you prefer.

Red cabbage dye

Red cabbage creates natural dyes that produce beautiful shades of blue, purple and even green. I collect the cabbage scraps and store them in the freezer until they are ready to use. Before freezing them, I chop the red cabbage into small pieces as this gives a better surface area for the dye. This is a really fun activity as you can also change the pH of the dye to create different colours using either vinegar or bicarbonate of soda.

> Soya milk (this is optional, but it is a natural mordant and will make the dye more stable)
> Red cabbage scraps to fill a large pan
> Natural fabric that is light in colour (cotton or linen are great options)

1 Prepare your fabric by washing it with a mild detergent – this allows the dye to adhere better to the fibres of the material.

2 If using soya milk you will need to weigh the fabric to determine how much you need. Measure out the mordant to weigh 10–15% of the fabric weight and dissolve in a large pan of water – just enough to submerge the fabric.

3 Bring the liquid to the boil and submerge the fabric into it. Simmer for about an hour and then rinse and set aside. This step is optional but will help the longevity of the dye.

4 Cut the cabbage into small pieces and place in a large pan.

5 Cover with water and bring to the boil, then reduce and let it

simmer for 1–2 hours. The longer you let it simmer the more concentrated your dye will become. You'll notice the water turning a deep purplish-blue colour.

6 Once the dye has reached an intensity that you like, remove the pan from the heat and allow it to cool a little.

7 Strain the liquid through a colander to remove the cabbage pieces.

8 You now have the option to change the colours of the dyes. For more purple hues, add a small amount of vinegar or a sprinkle of citric acid to the dye bath. For more green hues, add a small amount of bicarbonate of soda or soda crystals to the dye bath. You can have lots of fun with this process, and children absolutely love this step too because it is like a little magic trick.

9 Wet your pre-washed fabric thoroughly with water. This helps the dye to penetrate the fabric more evenly. Submerge the wet fabric into the dye bath, ensuring it is fully covered and immersed in the liquid.

10 Bring the dye bath back to the heat and let the fabric simmer for about an hour, stirring occasionally to make sure the colour penetrates the material evenly.

11 Take the pan off the heat. Leave the fabric in the dye bath for as long as you wish until the dye reaches a colour that you like.

12 Remove the fabric from the dye bath and rinse under cold water until the water runs clear.

13 Wash the dyed fabric with a mild detergent and hang to dry, out of direct sunlight, which can cause the colour to fade.

Tips:
The final colour of this dye will vary depending on the fabric type, amount of red cabbage used and how long you dye the material for. Natural dyes can be less colourfast than synthetic dyes, so wash dyed fabrics separately in cold water to prevent fading.

LOW-WASTE WATER ACTIVITIES

In the height of summer, I love to create some low-waste water activities that don't involve using the hosepipe for hours or making water balloons. These activities save water and will also keep your little ones entertained for hours.

Frozen flowers

This involves a lovely little walk in nature or the garden and lots of discussions about flowers too.

All you need to do is go and collect some of your favourite little flowers – I always choose the ones that are in abundance so that I leave plenty more for the pollinators, like daisies or rose petals. Flowers that have already been pollinated are great too.

> Flowers of your choice
> Ice-cube tray
> Boiled and cooled water

1. Once you have gathered all your flowers, bring them into the kitchen and pop them into ice-cube trays. Fill up the ice-cube trays with water that has been boiled and cooled (this makes the water less cloudy when they are frozen).

2. Pop the ice-cube trays in the freezer.

3. When they are frozen, take them out and let your children play with them. We have a little play kitchen area in the garden, so I pop them in the 'sink' area and let the children do what they like with them.

It's a good idea to choose edible flowers, as with young children we never know what they are going to do! Some edible flowers that work well with this activity are:

- Daisies
- Rose petals
- Nasturtiums
- Calendula
- Violas and pansies
- Chive blossoms
- Lavender
- Herbs
- Dandelions

Iceberg excavations

This is another lovely low-waste play activity and such an easy thing to do. It just takes a little planning the night before.

Gather up your children's toys (I tend to choose dinosaurs) and place them into muffin trays or silicone bags with water. Pop them into the freezer to freeze overnight so that they are set solid by the morning. Then add them to a play station or large bowl with some tools so that your children can 'excavate' the dinosaurs!

It is important here that you choose toys that won't be damaged in the freezing and thawing process. Wooden toys would not be ideal here as this would impact the quality of the toy. We have a collection of dinosaurs that we were able to buy second-hand that are perfect for this.

... FAMILY FUN 235

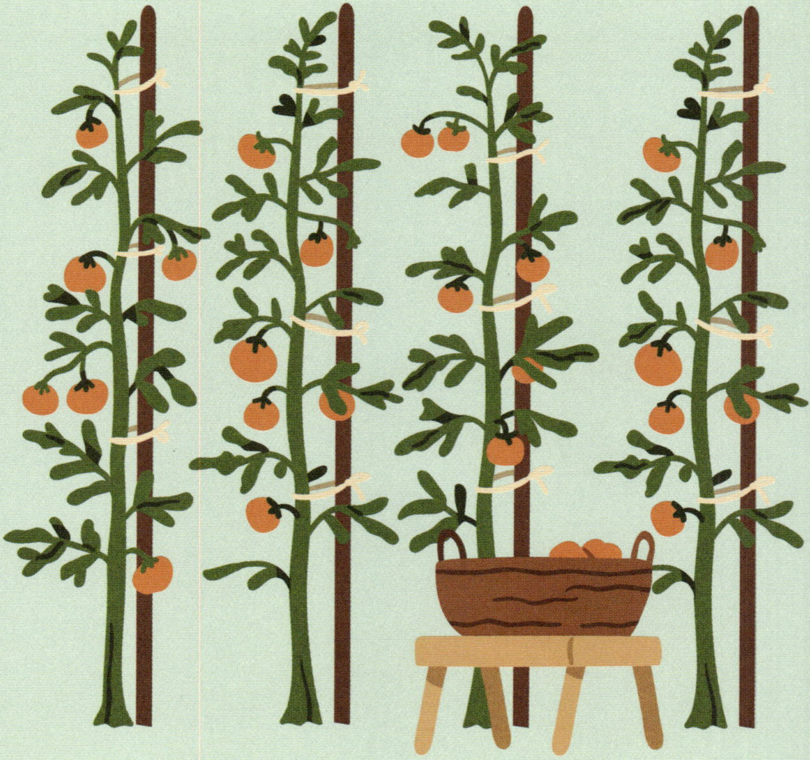

MY PLASTIC-FREE GARDEN

PLASTIC-FREE IN THE GARDEN

The production of plastic has had a huge impact on the gardening industry. You only need to pop to your local garden centre to see that there is a plastic option (usually disposable) for nearly every job in the garden. I find it's quite at odds with the objective of gardening and try to use natural materials wherever possible. I am merely an amateur gardener, but here are some of my favourite plastic-free and low-waste practices that I have adopted.

Avoid pesticides in your garden

To protect all of the pollinators that you introduce to your garden, it's so important to avoid pesticides. Pesticides are harmful to bees, butterflies and other pollinators; they can kill these insects directly or damage their immune systems, making them more susceptible to disease.

You can avoid pesticides in your garden by planting a variety of plants to create a balanced ecosystem. This will help attract a variety of insects and predators that naturally control pest populations. Introducing companion plants can also help manage pests in your garden. An example of this is I always try to plant nasturtiums around my garden as they attract aphids and white flies. By planting your nasturtium near more valuable crops that are usually affected by these insects, they tempt them away from the main crops and onto the companion plant. If this is something that you're interested in there is so much material available and I encourage you to research more into it.

... GARDEN

Use mulch to retain soil moisture and reduce the need for watering

Mulching is a great way to improve the health of your soil, conserve moisture and suppress weeds all at the same time. It's a really simple way to help your garden stay healthy and happy throughout summer and you can do this with lots of organic materials.

You can use wood chips, bark, straw, grass clippings, leaves, compost and even pine needles to mulch your garden. This organic mulch will decompose over time and improve soil quality.

1 Use a rake or your hands to spread the mulch evenly around your plants. Make sure to leave a gap around the base of trees and shrubs to prevent moisture build-up and potential rot. The gap should be about 5–8cm from the base of your trees and shrubs.

2 Periodically rake the mulch to prevent compaction. This can reduce air and water penetration to the soil.

The best time to mulch is early spring after the soil has warmed up. Mulching in autumn can also protect plants from extreme winter temperatures.

Seedling pots

I make a range of seedling pots to grow my seedlings. They are a great way to plant your seeds and cultivate your garden without all of the plastic.

Newspaper seedling pots

These pots are really easy to make and can be made in various shapes and sizes. All you need are strips of newspaper and a jar to use as a mould.

> Newspaper sheets (avoid using glossy pages)
> Scissors
> A small cylindrical object (like a glass jar) for shaping
> Compostable tape (optional)

1. Cut the newspaper sheets into strips approximately 12cm wide and as long as the newspaper sheet.

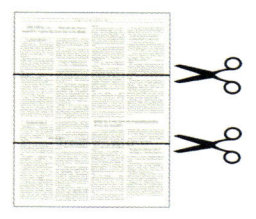

2. Gather the strips so that they are 2–3 ply and lay them flat on the table.

3 Place the cylindrical object at the corner of the strip.

4 Roll the newspaper around the cylinder, leaving 2–5cm extending beyond the bottom of the cylinder.

5 Tuck and fold the excess newspaper at the bottom of the open end of the cylinder. This will form the base of your pot. Gently press the bottom of the pot to the table to secure the folds and keep the bottom intact.

6 Carefully slide the cylinder out from the newspaper tube – you will have now made yourself a seedling pot.

7 If the bottom of your pot doesn't feel secure enough, use a little piece of compostable tape to secure it a little more.

8 Fill your pot with coconut peat-free compost (which I find is most widely available plastic-free) and plant your seeds.

Using newspaper pots is not only cost-effective but also helps reduce waste by repurposing old newspapers. It's a great way to start your seedlings without needing to purchase plastic pots.

Toilet roll seedling pots

Making seedling pots from toilet paper rolls is another eco-friendly gardening solution that means you can avoid the cheap, plastic seedling pots.

> Toilet roll inner tubes
> Scissors
> Compostable tape (optional)

1. Using scissors, make four evenly spaced vertical cuts about 2cm long at one end of the toilet roll inner tube. These cuts will form flaps that will be the base of the pot.
2. Fold each of the cut flaps inwards, overlapping slightly to create a closed bottom for the pot and press down firmly to secure the flaps.
3. If you want to make the bottom of the pots a little more secure, you can add a little compostable tape to secure them.
4. Fill your pot with coconut peat-free compost and plant your seeds.

Now I've discussed seedling pots I thought it might be a good time to talk about my favourite thing to grow in the garden – loofahs! Did you know that you can grow your very own plant-based sponge in the garden? I've been growing them for quite a few years now and here are my top tips for starting them off (which may be particularly handy if you live in a cooler climate).

1. Sow your seeds in early spring in a warm spot in your house that gets lots of direct sunlight. To speed up the process of germination you can soak your seeds before germination: place them in a damp towel like a napkin and pop them in a warm cupboard in a sealed container (this will keep the seeds moist).

2. When your seeds have sprouted you can pop them in some compost to grow.

3. The seedlings will grow quite quickly and you will need to pot them on. I tend to recycle old planting pots for this job – just make sure they are washed and clean before you use them.

4. When the weather has warmed up and you feel that it is safe to keep them in a greenhouse, you can now pop them in there.

5. Provide support for the plants much like you would do with a cucumber plant.

6. Luffa plants like water but do not like to be saturated so be sure not to overwater.

7. Leave the fruits to mature for use as sponges – this will take the whole summer so be patient.

8. Harvest the fruits and allow to dry before peeling your loofah sponge ready for use.

Saving seeds
Saving seeds from plants allows you to save money, maintain genetic diversity and grow plants from your favourite varieties. Wait until the seeds are fully mature on the plant. This varies by plant type and species. Seeds are typically ready when they have changed colour and become hard.

For fleshy fruits like tomatoes and strawberries
Remove the mature fruits from the plant. Scoop out the seeds along with the pulp of the fruit and place in a container. Add some water to the container and allow the fruit to ferment for a couple of days (this will help break down the pulp). After fermentation, rinse and strain your seeds, then spread them out on dry reusable kitchen towel or a screen and allow to dry in a dry, well-ventilated area.

For dry seed pods like peas and beans
Allow the pods to dry on the plant until they turn brown and begin to crack open. Harvest the pods before they crack open fully to prevent seed loss, then shell the pods to remove the seeds.

For flowers with seed heads like sunflowers and poppies
Allow the flowers to fully mature and dry on the plant. Cut the seed heads from the plant once they are dry but before they begin to scatter. Rub or shake the seed heads to release the seeds.

For small seeds like lettuce and herbs
Allow the seeds to mature on the plant until they are dry and easily detachable. Cut the seed heads or pods and collect them in a container, then crush or rub the seed heads to release the seeds.

Drying seeds
After harvesting, spread the seeds out on dry reusable kitchen towel or a screen in a warm, dry place with good air circulation. Allow the seeds to dry completely for about 1–2 weeks. Stir or turn them occasionally to ensure even drying.

Storing seeds
Once dry, store the seeds in labelled envelopes or small containers. Store seeds in a cool, dry place away from direct sunlight. Properly stored seeds can remain viable for several years.

DIY plant food

I like to make plant food from things that I would normally throw in the compost. It's a great way to repurpose and reuse something out of nothing. There are three types of plant food that I usually make.

Nettles

Nettles can be used to create a nutrient-rich plant food or fertiliser.

> Fresh nettles
> Gloves (to avoid stings)
> Large container (bucket or barrel)
> Water

1. Gather fresh nettles. Choose healthy plants without flowers, as younger plants are richer in nutrients. Use gloves and take care to avoid being stung.

2. Fill a large container with water. Rainwater is ideal but tap water can also be used.

3. Add the nettles to the water. You can crush or chop them up to increase the surface area and help release nutrients.

4. Cover the container loosely to allow airflow and prevent pests from getting in.

5. Place the container in a sunny spot or a warm area for 1–3 weeks. During this time, fermentation will occur, breaking down the nettles and releasing nutrients into the water.

6 After 1–3 weeks, strain the liquid through a sieve or cloth to remove the solid plant material. The liquid you've collected is now nettle plant food.

7 Dilute the nettle plant food with water before using by mixing 1 part nettle liquid with 10 parts water.

Use the diluted nettle plant food to water your plants. It can be applied to both indoor and outdoor plants, vegetables, flowers and trees.

Coffee grounds
Using coffee grounds as plant food is a great way to use up your old coffee grounds and give those acid-loving plants in your garden a feed.

> Used coffee grounds
> Large container
> Water

1 Add the coffee grounds to a large container – they will need to be diluted to a ratio of 1 part coffee grounds and 10 parts water.

2 Leave for 1–2 days.

3 Use the water to water your acid loving plants.

- Azaleas
- Camellias
- Rhododendrons
- Hydrangeas
- Roses
- Magnolias

Coffee grounds improve soil health and make a great plant food for your plants, but it is important not to overuse them. I would advise watering your plants with them monthly to enjoy the nutrient boost.

Banana peel tea

Banana peels are another great natural fertiliser as they have a high potassium content, as well as other nutrients like phosphorous and calcium. The easiest way to use banana peels to water your plants is to make a banana peel tea with them.

> Banana peels
> Large container
> Water

1 Add the banana peels to a large container, cover with water and allow to ferment for a couple of days. They would be best stored in a bucket in the garden with a lid or something similar. A jar is a good idea if you are making small quantities, but it's important to leave the lid loose so that pressure does not build up from the fermenting of the bananas.

2 After a couple of days, the liquid will be ready to use to water your plants. Drain the banana skins out of the mixture and compost them.

3 Plants that enjoy banana tea:

- Roses
- Hydrangeas
- Dahlias

- Geraniums
- Tomatoes
- Peppers
- Cucumbers
- Azaleas
- Camellias

Banana peels make a great plant food for your plants, but it is important not to overuse them. I would advise watering your plants with them monthly to enjoy the nutrient boost.

RESOURCES

Here is a list of resources to help you find many of the products and suppliers that I have discussed throughout the book. It isn't an exhaustive list and I recommend that you also do your own research. I try to focus on small, eco-friendly businesses and I avoid large corporations where possible.

As many of you will know, I own my own eco-friendly refill and home essentials shop – it's called The Natural Living Shop and it's located in the heart of Lancashire. You are always welcome to visit, or you can purchase most of the products that I mention in the book online.

The Natural Living Shop
Cedar Farm
Mawdesley
L40 3SY
thenaturallivingshop.co.uk

KITCHEN

Washing up liquid / spray
minimlrefills.co.uk
blueland.com
neatclean.com

Dish brushes
ecoliving.co.uk
notoxlife.com

Cellulose sponges
memotherearthbrand.com
ecoliving.co.uk

Pot scrapers
bambooswitch.com
ecoliving.co.uk

Bicarbonate of soda
ecoliving.co.uk

Wood balm
oakdalebees.co.uk
ardentgoods.co

Walnut oil
Local supermarket

Glass chopping board
lakeland.co.uk

Reusable kitchen towels
honeyco.uk
marleysmonsters.com

Swedish dishcloths
wildandstone.com
theseepcompany.com
threebluebirds.com

Silicone straws
thesiliconestrawcompany.co.uk
gosili.com

Stainless-steel straws
ecoliving.co.uk
packagefreeshop.com

Glass straws
glassstraw.co.uk

Bamboo straws
wildandstone.com

Coffee cups / travel flasks
kleankanteen.com
black-blum.com
stojo.co

Insulated water bottles
chillys.com
kleankanteen.com

Stainless-steel food containers
black-blum.com
moonmoon.co.uk
asliceofgreen.com
elephantbox.co.uk

Silicone freezer bags
moonmoon.co.uk
stasherbag.com

Silicone baking mats
greenislandco.com
ecoliving.co.uk

Tea bags
yogitea.com
hampsteadtea.com

BATHROOM

Sweet almond oil
thenaturallivingshop.co.uk

Vegetable glycerine
thenaturallivingshop.co.uk

Soap
friendlysoap.co.uk
puresoapworks-us.com
ecoliving.co.uk

Soap saver pouch
notoxlife.com
thenaturallivingshop.co.uk

Essential oils
valleymist.co.uk
nikura.com
nealsyardremedies.com

Preservative Eco
thenaturallivingshop.co.uk

Konjac sponge
moonie.co.uk
memotherearthbrand.com

Loofah
ecobathlondon.com

Pumice stone
ecobathlondon.com

Cellulose face and duck sponge
thenaturallivingshop.co.uk

Shampoo bars / conditioner bars
thenaturallivingshop.co.uk
ecoliving.co.uk
faithinnature.co.uk
vioribeauty.com

Toothpaste in glass jars and aluminium tubes
benandanna.uk

Toothpaste tabs
ecoliving.co.uk

Electric toothbrushes
trysuri.com

Toothbrushes
thetruthbrush.com

Dental floss
ecoliving.co.uk
behuppy.com

Coconut oil
Local supermarket

Tongue scraper
goldricknaturalliving.co.uk
behuppy.com

Cornflour / cornstarch
Local supermarket

Reusable period underwear
wuka.co.uk
saalt.com

Reusable mentrual cups
allmatters.com
wearemooncup.com

Skincare ingredients
thenaturallivingshop.co.uk
acalaonline.com

Citric acid
ecoliving.co.uk

CLEANING

Isopropyl (rubbing) alcohol
Widely available online

Vinegar
minimlrefills.co.uk

Sodium percarbonate (natural bleach)
thenaturallivingshop.co.uk
ecoliving.co.uk

Eco-friendly cleaning products / closed-loop systems
neatclean.com
minimlrefills.co.uk
blueland.com
branchbasics.com
notoxlife.com

Eco Egg
ecoegg.com

Dryer balls
packagefreeshop.com
thenaturallivingshop.co.uk

Stain remover bars
chateaudusavon.com

PICTURE CREDITS

All photographs © Alejandra Sinclair, with the following exceptions:
Pages 29, 36, 47, 107, 121, 143, 155, 171, 217, 220 and 232: Shutterstock.com
Page 239: Moodboard Stock Photography / Alamy Stock Photo

All illustrations Shutterstock.com with the exception of pages 197–200, 202–205 and 241–243: Liane Payne © HarperCollins*Publishers* 2025

HarperCollins*Publishers*
1 London Bridge Street
London SE1 9GF

www.harpercollins.co.uk

HarperCollins*Publishers*
Macken House, 39/40 Mayor Street Upper
Dublin 1, D01 C9W8, Ireland

First published by HarperCollins*Publishers* 2025

10 9 8 7 6 5 4 3 2 1

Text © Kate Jones 2025

Kate Jones asserts the moral right to be identified as the author of this work

A catalogue record of this book is available from the British Library

ISBN 978-0-00-871574-8

Printed and bound at RR Donnelley, China

All rights reserved. No part of this publication may be reproduced, stored in a retrieval system, or transmitted, in any form or by any means, electronic, mechanical, photocopying, recording or otherwise, without the prior written permission of the publishers.

Without limiting the author's and publisher's exclusive rights, any unauthorised use of this publication to train generative artificial intelligence (AI) technologies is expressly prohibited. HarperCollins also exercise their rights under Article 4(3) of the Digital Single Market Directive 2019/790 and expressly reserve this publication from the text and data mining exception.

MIX
Paper | Supporting responsible forestry
FSC™ C007454
www.fsc.org

This book is produced from FSC™ certified paper and other controlled sources to ensure responsible forest management.

For more information visit: www.harpercollins.co.uk/green